Application of the Neutral Zone in Prosthodontics

Application of the Neutral Zone in Prosthodontics

Joseph J. Massad, DDS

Private Practice, Tulsa, OK, USA
Associate Professor, Department of Prosthodontics, University of Tennessee Health Center,
School of Dentistry, Memphis, TN, USA
Adjunct Associate Professor, Department of Prosthodontics and Operative Dentistry,
Tufts University School of Dental Medicine, Boston, MA, USA
Adjunct Associate Professor, Department of Comprehensive Dentistry,
University of Texas Health Science Center, School of Dentistry, San Antonio, TX, USA
Adjunct Associate Professor, Department of Restorative Dentistry,
Loma Linda University School of Dentistry, Loma Linda, CA, USA
Clinical Assistant Professor, University of Oklahoma College of Dentistry,
Oklahoma City, OK, USA

David R. Cagna, DMD, MS

Professor, Department of Prosthodontics
Associate Dean, Postgraduate Affairs
Director, Advanced Prosthodontics Program
University of Tennessee Health Science Center, College of Dentistry, Memphis, TN, USA
Diplomate & Director, American Board of Prosthodontics
Fellow, American College of Prosthodontists

Charles J. Goodacre, DDS, MSD

Distinguished Professor, Department of Restorative Dentistry
Loma Linda University School of Dentistry, Loma Linda, CA, USA
Diplomate and Past-President, American Board of Prosthodontics

Russell A. Wicks, DDS, MS

Professor and Chairman, Department of Prosthodontics
University of Tennessee Health Science Center, College of Dentistry, Memphis, TN, USA

Swati A. Ahuja, BDS, MDS

Adjunct Assistant Professor, Department of Prosthodontics
University of Tennessee Health Science Center, College of Dentistry, Memphis, TN, USA
Prosthodontic Consultant, Lutheran Medical Center, New York City, NY, USA

WILEY Blackwell

The right of Joseph J. Massad, David R. Cagna, Charles J. Goodacre, Russell A. Wicks and Swati A. Ahuja to be identified as the authors of this work has been asserted in accordance with law.

Registered Offices
John Wiley & Sons, Inc., 111 River Street, Hoboken, NJ 07030, USA
John Wiley & Sons Ltd, The Atrium, Southern Gate, Chichester, West Sussex, PO19 8SQ, UK

Editorial Office
111 River Street, Hoboken, NJ 07030, USA

For details of our global editorial offices, customer services, and more information about Wiley products visit us at www.wiley.com.

Wiley also publishes its books in a variety of electronic formats and by print-on-demand. Some content that appears in standard print versions of this book may not be available in other formats.

Limit of Liability/Disclaimer of Warranty
The contents of this work are intended to further general scientific research, understanding, and discussion only and are not intended and should not be relied upon as recommending or promoting scientific method, diagnosis, or treatment by physicians for any particular patient. In view of ongoing research, equipment modifications, changes in governmental regulations, and the constant flow of information relating to the use of medicines, equipment, and devices, the reader is urged to review and evaluate the information provided in the package insert or instructions for each medicine, equipment, or device for, among other things, any changes in the instructions or indication of usage and for added warnings and precautions. While the publisher and authors have used their best efforts in preparing this work, they make no representations or warranties with respect to the accuracy or completeness of the contents of this work and specifically disclaim all warranties, including without limitation any implied warranties of merchantability or fitness for a particular purpose. No warranty may be created or extended by sales representatives, written sales materials or promotional statements for this work. The fact that an organization, website, or product is referred to in this work as a citation and/or potential source of further information does not mean that the publisher and authors endorse the information or services the organization, website, or product may provide or recommendations it may make. This work is sold with the understanding that the publisher is not engaged in rendering professional services. The advice and strategies contained herein may not be suitable for your situation. You should consult with a specialist where appropriate. Further, readers should be aware that websites listed in this work may have changed or disappeared between when this work was written and when it is read. Neither the publisher nor authors shall be liable for any loss of profit or any other commercial damages, including but not limited to special, incidental, consequential, or other damages.

Library of Congress Cataloging-in-Publication data applied for

ISBN: 9781119158141 [Hardback]

Cover Design: Wiley
Cover Image: (Background) © caracterdesign/Gettyimages; (Faces) Courtesy of Todd Heilmann

Contents

Foreword

The Need to Understand Edentulous Patients

With persistent efforts towards improved oral care, the world is experiencing a decline in the number of edentulous individuals. Despite this, the need for complete denture treatment is still in demand. In fact, dentistry continues to offer innovative denture services, such as computer-assisted design, computer-assisted manufacturing, better-controlled resin processing methodologies, and new materials. Recent developments in denture therapy largely reflect the evolving mindset of patients around the globe. Dentistry has come a long way from the days of vulcanite and ivory teeth. The evolving thoughts, perceptions, and expectations of patients are underpinning denture evolution and denture service delivery.

A major shift in demographics has fueled this evolution. Developing countries have witnessed a steep increase in life expectancy. In Japan life expectancy is now 83.7 years, and in the United States 79.3 years. The same is true in developing nations like India, where life expectancy is now 68.3 years. These changes are related to improved medical care. Individuals living longer and previous dental neglect, combine to create a cohort possessing complex dental problems that require specialized prosthodontic management.

Improved economic stability, resulting from monetary and insurance reform, also influences changing attitudes on dental rehabilitation with complete dentures. Again this trend is clearly evident in developed nations, but also significant in developing countries. More than ever before, patients are electing to invest in enhanced denture services.

Continuous movement of populations across borders has heightened awareness of the benefits of oral health worldwide. Dentists regularly encounter patients well versed in the importance of good oral health and treatment necessary for its maintenance, to include the full range of prosthodontic services. Dentures are no longer seen as a "taboo." Patients appreciate oral health as a gateway to improved general health, and insist on the optimal replacement of missing teeth.

The improved lifestyles of modern populations catalyze the demand for optimal denture services. The desire to elevate personal image and social acceptance, regardless of age, drives patients seeking esthetic dental rehabilitation. Lost facial support leading to an aged appearance, encourage patients to seek esthetic improvements through quality denture services. Likewise, replacing old complete dentures that fail to provide esthetic advantage occurs more frequently today than ever before in dentistry.

Changing dietary patterns and food selections, with greater emphasis on foods requiring efficient masticatory function, lead patients to appreciate well-functioning prosthetic replacement of missing teeth and to seek high-quality denture service. With edentulism and/or inadequate complete dentures, patients are at risk of suboptimal nutritional intake due to compromised masticatory ability. This too is a common complaint from patients seeking high-quality denture services.

In the face of this increased demand, it is important to revisit the classic dictum put forward by Dr. Muller M. DeVan[1] so many years ago, "*The dentist should meet the mind of the patient before he meets the mouth of the patient.*" Unless we understand our patient, his motivations, and the road he traveled to edentulism, any dental rehabilitative effort will be compromised at best. The patient may be elderly, reporting to the dentist as a result of tooth loss over several decades of organized personal hygiene neglect. He may be middle aged and suffering from masticatory inefficiency as a result of anodontia, hypodontia, ectodermal dysplasia, or some similar disorders, or due to traumatic tooth loss. A thorough understanding of the motivations driving these dramatically different patients may provide insight to both mindset and expectations. Unrealistic expectations are all-too-often therapeutically insurmountable, requiring that the patient be made aware of treatment and prosthesis limitations. Frequently, older patients lack tolerance for, and compliance with, the long appointment times required for optimal prosthodontic treatment. When successfully detected during the first patient interview, this important consideration will likely influence appropriate treatment selection.

In summary, it is paramount to "*meet the mind*" of edentulous patients so that rehabilitative dental therapy can be optimized. Several considerations introduced in these few introductory paragraphs, and further detailed throughout this important text, will aid enthusiastic and meticulous dentists in greater appreciation of edentulous patients in order to offer sound solutions in the management of their concerns.

Prof. (Dr.) Mahesh Verma
Director – Principal, Maulana Azad Institute of Dental Sciences, MAMC Complex, BSZ Marg, New Delhi, India

Dr. Aditi Nanda
Senior Research Associate, Department of Prosthodontics, Maulana Azad Institute of Dental Sciences, MAMC Complex, BSZ Marg,
New Delhi, India

Reference

1 DeVan MM. Methods of procedure in a diagnostic service to the edentulous patient. *J Am Dent Assoc* 1942;29:1981–1990.

Preface

The concept of the neutral zone is by no means original and was discussed in 1933 in a textbook titled *The Principles of Complete Dentures*, authored by Sir Wilfred Fish. In 1973, Dr. Victor Beresin and Dr. Frank Schiesser published a textbook titled *The Neutral Zone in Complete Dentures*. The neutral zone concept was initially intended for edentulous patients; however, in 1978, Beresin and Schiesser published a second edition titled *The Neutral Zone in Complete and Partial Dentures*.

Even though complete dentures are not ideal replacements for natural dentition, they should not be noticeable or feel like a foreign object in the patient's mouth. Incorporating the actions of the surrounding muscles of facial expression, speech, and mastication is often overlooked in the fabrication of maxillary and mandibular complete dentures. All oral functions that include chewing, swallowing, speaking, laughing, and sucking, involve the harmonious action of the lips, cheeks, tongue, and the floor of the mouth. These actions have an influence on prosthetic design and can be recorded by a functional method. Failure to acknowledge these functions can affect tooth positions, border extensions, the occlusal plane location, and the contours of the polished surface, which may result in unstable and unsatisfactory prostheses. The concept of the neutral zone takes into account the neuromuscular functions that contribute to denture stability. This book will discuss and illustrate a step-by-step method to identify and record the neuromuscular actions that help to define

appropriate tooth positions and develop cameo surface contours which feel more normal to the patient. This method has been expanded to include the dentate patient preparing for immediate dentures, implant-supported overdentures, and fixed complete dentures (hybrid implant prostheses). Complete dentures fabricated by the methods described, can become a guide for optimal placement of implants within the confines of the prosthesis contours.

Successful complete denture therapy is often a considerable challenge for the less-experienced practitioner, so many dentists choose to limit, or not offer, this service in their practice. The number one denture problem reported by dentists globally has been fit and stability, followed closely by occlusal disharmony and compromised esthetics. This book is similar to others on this subject in that it will cover all phases of complete denture records and fabrication. It reviews a step-by-step assessment and examination protocol, designed to deliver an accurate diagnosis and prognosis prior to committing to treatment. It also describes a very predictable "impressioning" procedure that can be accomplished in a single appointment with a level of accuracy that is similar to, or better than, conventional methods. It discusses the severely vertically closed patient and the resolution of this condition, and describes the techniques of making maxillo-mandibular jaw relationship records to accommodate optimal treatment results.

The problem of esthetics, one of the most critical issues plaguing the dental practitioner, can be avoided. As esthetics is on most patients' minds today, we have dedicated a portion of the first chapter to the identification and hopefully the elimination of any unrealistic patient demands. In this textbook, we have utilized the concept of "anticipating failure in order to avoid it." Lack of knowledge and failure to recognize the patient's desires and needs can and will have a disastrous effect on the prognosis of any prostheses. However, if we understand human nature and ask the right questions of each of our patients, then it becomes much easier to understand their actual requirements and allows them to be part of the process in building the esthetic result.

The primary objective of this book is to describe current procedures in the fabrication of complete dentures by blending multiple clinical procedures and philosophies to create a contemporary recipe for optimal outcomes. The intent is to identify fundamental applications that can be related to various prosthodontic procedures practiced today. Another goal is to empower the reader with additional knowledge, confidence, and practical applications in the provision of prosthodontic services.

To begin the journey, I would like to thank all those who have assisted me in becoming a more astute, compassionate and learned practitioner; particularly my mentors Drs. Frank Schiesser, Kenneth Rudd, Thomas Shipmon, Lindsay Pankey, and John Frush. I am very grateful to have highly respected co-authors, Drs. David Cagna, Charles Goodacre, Russell Wicks, and Swati Ahuja, and contributing author Mahesh Verma. A special thank you to Dr. Ahuja for compiling all the information from all the authors to complete this manuscript. I want to add my sincere thanks to my multiple reviewers for their suggestions as to the content of this manuscript. They are Drs. Mahesh Verma, Tony Daher, David Little, William Davis, Mostafa ElSherif, Richard June, William Lobel, Samuel Strong, and Joseph Thornton. It is also very important to me to thank Mr. Todd Heilmann for his expertise in taking and preparing all the photographs and illustrations.

A thank you to Kenneth Waldo, Ron Johnston, Eric Newnum, Craig Nelson, and Zarko Danilov, my dental prosthetic technicians, who worked tirelessly in preparing actual patient cases so that I could demonstrate vital aspects of this manuscript. Also a thank you to William Knowles for his engineering of the dental devices we used in treating our patients. I want to thank Dr. John Gordon who invited me to Jamaica to begin writing this manuscript in isolation while his sister, Glass, typed every word as I dictated for four long days.

I dedicate this textbook to my lovely wife Darlene, and my wonderful children Jolene, Jordan, Joshua, Jodain, and Joslyn.

Joseph J. Massad

About the Companion Website

This book is accompanied by a companion website:

www.wiley.com/go/massad/neutral

The website includes:

- Video clips
- Student handouts for download

Your password for the site is **ghx19cb354e.**

Instructors can also gain access to a companion website with the above materials and instructional PowerPoints, which are for faculty use only and should not be distributed to students. To access this site, please go to the book's page on wiley.com and navigate to the Instructor Site; you can then register your information to gain access.

1

Assessment of Edentulous Patients

Introduction

A critical and somewhat perplexing aspect of the management of the edentulous condition is the prediction of therapeutic outcomes and patient satisfaction. The most fundamental factor determining a precise prognosis is a thorough and accurate pretreatment examination [1–3]. Even though patients may receive the best therapy, the treatment will fail if underlying conditions remain undiagnosed.

This chapter reviews a method for the pretreatment evaluation of edentulous patients and existing prostheses to arrive at a sound understanding of factors that will affect therapy and the probability that the treatment's objective can be achieved. Using appropriate assessment tools, the practitioner can better determine if the patient's expectations can be met.

Much has been published in the dental literature regarding anatomic [4, 5] and psychological variations [6, 7] in edentulous patients. Before considering management of these challenging patients, objectives include thorough examination, diagnosis of existing conditions, consideration of available therapy, and assessment of the prognosis of each available treatment option [1, 2]. Both subjective and objective patient factors must be taken into consideration [1]. A rational stepwise pretreatment protocol will help to prevent critical diagnostic information from being overlooked. Detailed documentation of findings is essential from a dento-legal standpoint.

The pretreatment protocol provided is relatively easy to follow, quick to perform, and easy to reproduce. It yields summary findings that correspond with specific prognostic conclusions. The protocol is divided into: (i) patient interview; (ii) examination of existing facial characteristics; and (iii) examination of edentulous conditions, i.e., anatomic, morphologic, and muscular status.

The Patient Interview

Successful therapy is facilitated by the provider coming to know the patient, from both personal and logistical perspectives; this includes how the patient arrived in the practice. If the patient was referred, the referral source should be known and contacted, and the reason for the referral noted. If the patient arrived due to marketing of the practice, care must be taken to investigate if the patient's needs are consistent with therapy provided by the practitioner.

The initial patient interview permits the patient and the practitioner to know one another [8]. Quality time spent at the beginning sets the stage for an optimized patient-provider relationship. Both the physical and psychological status of the patient should be triaged during the first appointment [8]. Anticipation of communication problems and interception of commonly encountered interpersonal

Application of the Neutral Zone in Prosthodontics, First Edition. Joseph J. Massad, David R. Cagna, Charles J. Goodacre, Russell A. Wicks and Swati A. Ahuja.
© 2017 John Wiley & Sons, Inc. Published 2017 by John Wiley & Sons, Inc.
Companion website: www.wiley.com/go/massad/neutral

problems are frequently as important as clinical findings. Discerning the primary etiology of existing patient dissatisfaction is essential for breaking the cycle of unsuccessful treatment attempts. Complaints and expectations expressed by the patient, and treatment obstacles encountered by previous dentists, can provide a critical influence on the acceptance of the patient into the practice and the treatment offered.

Be aware that the pretreatment protocol provided might initially appear to consume an inordinate amount of time and effort. Some might say that this is financially unjustifiable. However, once understood and skillfully conducted, the protocol reduces overall management time, permits appreciation of the treatment rendered, and significantly contributes to overall therapeutic success.

Some patients may be fearful, nervous, or shy, and inadvertently fail to respond directly to questions. Recognition of these individuals early in the interview process is critical. In many cases, a dental auxiliary can better elicit patient responses than the practitioner. Obtaining honest and accurate patient responses will affect outcomes. The pretreatment protocol and associated electronic documentation presented incorporate data-gathering processes designed to elicit thorough, concentrated, and accurate answers from patients.

Patient Interview: Age

The patient's chronological age should be critically compared with general physical health and existing oral conditions. Older patients may be afflicted with poor neuro-muscular coordination [9, 10], suboptimal nutritional status [11, 12], diminished adaptability [9, 10], and salivary secretion (both quantity and quality) [11], and highly vulnerable denture-bearing tissues [10, 11]. These factors adversely influence aging edentulous patients' ability successfully to tolerate and function with conventional complete dentures, which should be discussed prior

to initiating treatment [8]. Analogies such as "when dentures move and there's limited saliva, the pink plastic acts like sand paper against your gums creating irritation" help patients to understand better the problems that they face.

Patient Interview: Attitude

Coming to appreciate patient attitude may be as simple as presenting nonleading questions and permitting the patient time to respond. Questions that may be used to gauge patient attitude include:

- *How are you feeling today?*
- *How was your experience with the previous dentist that treated you?*
- *What do you think about your current and previous dentures?*

Based on patient responses and ensuing discussions, qualifications of patient attitude as good, average, or poor may be made. Of course, additional questioning may be necessary to arrive at a reasonable determination.

Patient Interview: Expectations

If not thoroughly investigated prior to initiating treatment, patient expectations may not be apparent until problems unexpectedly emerge in the course of therapy, and the patient's demeanor begins to decline [9, 13]. Direct and specific questioning of the patient regarding expectations will permit documentation of responses and qualification of expectations as high, medium, low, or still unsure. Patients can also be asked the following questions to understand further the nature of their expectations:

- *What kind of improvement in appearance do you expect from your new dentures?* In response to this question, a 50-year-old patient may provide a picture of an 18-year-old celebrity stating, "I want my teeth to look like hers." This would indicate that the patient possesses unrealistic expectations. A subsequent patient may suggest, "I want perfect teeth," necessitating a better understanding of what is meant by "perfect."

- *What kind of improvement in chewing ability do you expect from new dentures?*
- *What kind of improvement in fit do you expect from the new dentures?*
- *How long do you expect new dentures to last?*
- *How often do you expect to return to the dentist for examinations and adjustments?*

The nature of the patient's desires and demands relative to proposed treatment must be considered by the practitioner within the context of his/her level of experience and expertise. If the patient expects more than the practitioner can comfortably provide, definitive treatment should not commence and referral to a more experienced colleague should be in order. Additionally, if the patient is unable to appreciate the limitations of the therapy offered, it is inappropriate to initiate treatment.

It is the responsibility of the practitioner to address unattainable expectations fairly and honestly, through frank discussion with the patient, communicating what can and cannot be accomplished with treatment; this is particularly true with complete denture therapy. Failure to address unrealistic expectations often leads to treatment failure and rapid deterioration of the patient-provider relationship. Patients that refuse to accept known limitations of therapy and express inflexibility in this regard are generally challenging to manage successfully. Not initiating definitive treatment for these individuals is ethically, professionally, and financially appropriate.

Patient Interview: Chief Complaint

Providing state-of-the-art treatment that does not manage the patient's main concerns may provide a level of personal satisfaction for the provider but is rarely successful in the long run. It is therefore important to: (i) request that patients specifically voice their greatest dental concern/concerns; (ii) document these chief concerns using the patients' exact words, and (iii) review the chief concern/concerns, as documented, with the patients to confirm accuracy [13].

Most dental patients are not familiar with professional and dental terminology. It is therefore important to ensure that the practitioner understand clearly the patient's chief concerns as expressed. Asking the following questions may permit a greater appreciation for the nature of the chief concerns:

- *Are your dentures loose?*
- *Can you eat most foods?*
- *Do your gums get sore?*
- *Do you have pain now?*
- *Are you happy with the appearance of your smile?*
- *Is there anything else that bothers you?*

Patient Interview: General Health

General health is a significant factor that can affect the overall success of dental therapy [9]. A thorough medical history questionnaire is an essential tool in pretreatment diagnosis. Patients with complicated medical conditions (e.g., uncontrolled diabetes, Parkinson's disease, Huntington's disease, Tourette's syndrome, other neuromuscular disorders, etc.) should be informed that these conditions may affect their ability to retain and function with conventional complete dentures [14]. Many systemic conditions (e.g., iron deficiency anemia, Sjogren's syndrome, pemphigus/pemphigoid, erythema multiforme, etc.) can adversely affect oral tissues, oral function, and in turn the success of complete denture therapy [14]. Obtaining information regarding current medication type and dosage is important, particularly because so many medications significantly contribute to xerostomia. Patients should be referred to their primary physicians for review of medical conditions or medications expected to affect dental therapy adversely.

Patient Interview: Complete Denture Experience

In order to assess patients' ability to wear removable prostheses and the apparent rate of alveolar bone resorption, they should report the number of years they have worn complete dentures [9, 13]. They should be questioned if the maxillary and mandibular dentures were fabricated at the same time or at different times. It is also important to note the reasons for tooth loss. As a general rule, longer durations of edentulism correspond to greater alveolar bone loss and increased complexity of treatment.

Patient Interview: Denture Remake Frequency

Information should be collected on the number of different complete dentures worn by the patient since loss of the natural teeth. The date of fabrication of the most recent complete dentures should be determined. Reasons for seeking new prostheses, both historically and currently, should be noted. The American Dental Association recommends that complete dentures be replaced every 5–7 years, or when they can no longer be worn comfortably [15]. Acquiring one new denture over the past 10 years is reasonable; two new dentures in 10 years may be justifiable, but three or more complete dentures within a 10-year period may indicate particularly challenging conditions or a challenging patient who is difficult to treat successfully.

Patient Interview: Patient Satisfaction

Satisfaction level with previous complete dentures is important diagnostic information [9]. Satisfaction should be qualified as successful, reasonably successful, or unsuccessful. The following specific questions should be asked:

- *Describe your satisfaction with previous dentures?*
- *What is your opinion regarding your smile with your existing dentures?*
- *Were you able to function with previous dentures?*
- *Did your dentures fit well in your mouth?*

Patient Interview: Photographs, Diagnostic Casts, and Radiographs

Photographs, diagnostic casts accurately mounted in an articulator, and radiographs are essential to complete the patient interview and information gathering. Properly composed photographs help to visualize smile symmetry, incisal display, lip support, size and form of edentulous ridges, and presence of undercuts. Mounted diagnostic casts present three-dimensional information on oral contours of the edentulous jaws, ridge relationships, and available restorative space.

Important objective diagnostic information is discernable with panoramic radiology. Relative alveolar height and resorptive patterns can be assessed. Hypertrophied tuberosities, pneumatized sinuses, and extruded ridge segments may be identified. Approximately 20% of edentulous patients present with radiographic signs of bone cysts, retained root tips, impacted teeth, and residual pathology [13, 16]. Incorporation of a properly made, diagnostic-quality panoramic radiograph early in the pretreatment protocol is essential in identifying these treatment concerns.

The Facial Analysis

Esthetic outcomes in modern dentistry are essential to perceived success [9]. Unfortunately, appreciating patient esthetic expectations and determining esthetic prognosis during initial assessment can be challenging. A detailed facial analysis involving patient interaction and acceptance is a critical element of the pretreatment protocol. Identification of dental midline asymmetries, lip irregularities, tooth and excess denture base displays, face shape, and vertical/horizontal residual ridge relationships influence both the treatment rendered and prognosis. Patient and dentist appreciation for these esthetic factors prior to initiation of treatment is best accomplished using carefully composed clinical photographs.

(a)

(b)

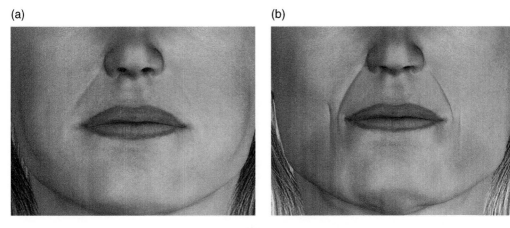

Figure 1.1 (a) A female patient with convex appearance; (b) a female patient with concave appearance.

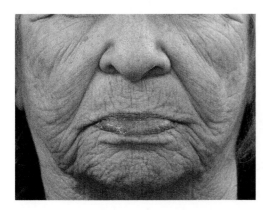

Figure 1.2 Patient demonstrating presence of deep wrinkles, nasolabial folds, poor muscle mass and tone.

Facial Analysis: Facial Tissue Tone

Aging and the loss of teeth correspond to deterioration of tonicity in facial tissues and masticatory muscles. Decreasing muscle mass alters the appearance of the face from relatively convex to concave (Figure 1.1a and b). Development of surface wrinkles, deep nasolabial folds, and concave cheek contour (Figure 1.2), are indicative of poor skin tone and underlying muscle mass. Digital palpation and patient history (e.g., complaint of reduced bite force) provide information on the tone and functional capacity of facial and masticatory muscles. When performing digital palpation, a thumb is placed near the commissures (Figure 1.3a) and the index and middle fingers on the opposite cheek surface (Figure 1.3b). The patient is asked to pucker the lips (Figure 1.3c) and then smile (Figure 1.3d). If these movements displace the fingers and the thumb, muscle tone is deemed adequate.

The tone of the oral and facial muscles following the loss of teeth may be near normal or subnormal but never normal [3]. The masticatory force and efficiency for complete denture wearers are therefore substantially reduced compared to those with natural dentitions [3]. The timing and sequence of tooth loss will affect muscle groups to varying degrees. If anterior teeth have been missing for some time, the muscles of facial expression will exhibit a poor tone. If posterior teeth have been missing for a long time, the muscles of mastication are more likely to exhibit a poor tone [3].

Adequate muscle tone contributes to denture stability. Patients with substantially poor muscle tone may find it difficult to stabilize complete dentures. Normal tension, tone, and placement of muscles in the absence of degenerative changes is ideal. However, muscle degeneration in edentulism is common.

(a)

(b)

(c)

(d)

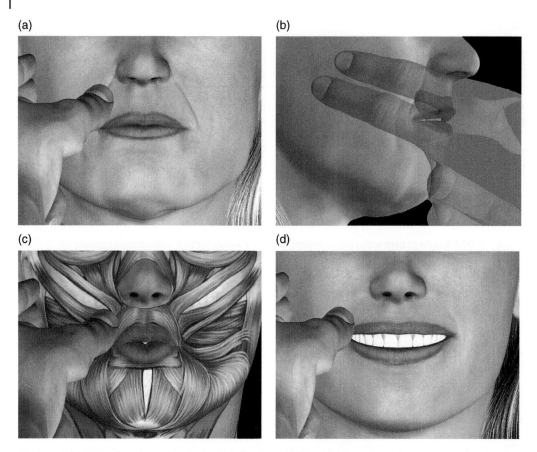

Figure 1.3 (a) Evaluation of muscle tone by digital palpation; (b) evaluation of muscle tone by placing the index finger and middle finger on the cheek; (c) evaluation of muscle tone by asking patient to pucker their lips; (d) evaluation of muscle tone by asking patient to smile.

An important function of complete dentures is to provide support for the muscles and soft tissues of the cheeks and lips. Denture flange borders and cameo surface contours should be developed to facilitate this support.

Facial Analysis: Tooth and Denture Base Display

Lip length and lip mobility affect tooth and soft tissue (denture base) display during both repose and smile. A long upper lip and reduced lip mobility during smile results in minimal maxillary tooth and gingival display (Figure 1.4a). A short upper lip and excessive lip mobility lead to maximum maxillary tooth display, particularly during full smile (Figure 1.4b). Tooth and denture base display of the existing prostheses during repose and full smile should be recorded to indicate no show, slight show, average show, or excess show. Recording this information helps to improve vertical denture tooth positioning in planned prostheses.

Facial Analysis: Midlines

The patient's maxillary denture midline should coincide with the facial midline. Deviations should be noted.

(a) (b)

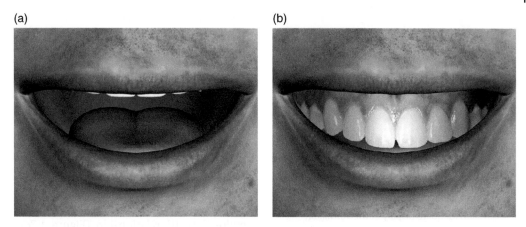

Figure 1.4 (a) Inadequate maxillary incisal display in smile; (b) excess maxillary incisal display during full smile in an old male patient.

(a) (b)

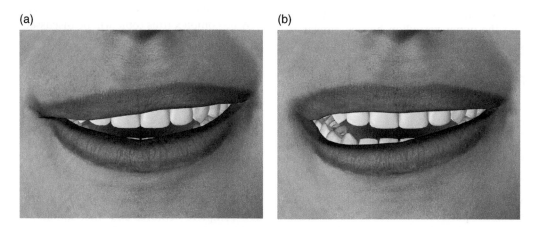

Figure 1.5 (a) Asymmetric movement of the upper lip; (b) asymmetric movement of the lower lip.

Facial Analysis: Lip Mobility

Symmetrical lip movement should be assessed during smile and full animation. Asymmetric lip movements should be classified as normal, slight, medium, or extreme. Photographs (e.g., repose and smile) and patient history (e.g., complaints of asymmetric tooth display) are valuable aids. Unilateral reduction of mobility (Figure 1.5a and b) and unilateral irregular contours should be noted (e.g., stroke or Bell's palsy) [9]. The position of anterior denture teeth and the cameo denture base contour can be manipulated to modify asymmetric lip positions and movements subtly, although complete correction may not be possible. Patients should be made aware of any lip asymmetries and the potential for corrective measures prior to initiating treatment. Referral for tissue fillers and plastic surgical procedures may be indicated.

Facial Analysis: Lip Dimension

Both upper and the lower lip dimension should be examined and classified as full, reduced, or minimal. Thin lips and vermillion borders become less visible with age (Figure 1.6). Labial inclination of maxillary anterior teeth can enhance the upper vermillion display. However, significant forward

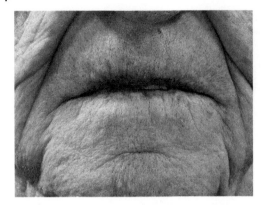

Figure 1.6 Inadequate display of the vermillion border of upper and the lower lip in a male patient.

positioning of maxillary incisal edges can thin the upper vermilion border, similar to stretching a flat rubber band, the greater it elongates the more it thins. Vermillion display can also be enhanced by tissue fillers.

Prosthetic Factors

Prosthetic Factors: Vertical Dimensions

With the current complete dentures in place, the patient's occlusal vertical dimension (OVD) and rest vertical dimension (RVD) should be compared. The RVD is recorded by marking a dot on the tip of the patient's nose and on the forward prominence of the chin [17]. The patient is asked to take a deep breath and relax. Once relaxed, a caliper is used to record the distance between dots (Figure 1.7a) [17]. It may help to have patients breath in and out several times, close their eyes as if to fall asleep / relax, and permit jaw muscles to relax in order to obtain RVD. This technique may be particularly helpful for patients who present with an apparently overreduced vertical posture. This measurement represents the patient's RVD or the physiological rest position. Next, the patient's existing OVD is recorded by having the patient occlude their denture teeth. Once again the caliper is

used to record the distance between the two dots (Figure 1.7b) [17, 18].

Subtracting the OVD from the RVD yields the dimension commonly known as freeway space. It is generally accepted that physiologically acceptable interocclusal distance (freeway space) for complete denture patients ranges from 2.0 to 4.0 mm [17]. Inadequate or excessive OVD / interocclusal distance can adversely affect the success of complete denture therapy [17, 18]. Inadequate OVD may result in mandibular overrotation with relative forward positioning, known as pseudo-Class III relationship (Figure 1.7c), compromising esthetics, masticatory efficiency, and denture stability [17, 18]. Management of patients with reduced OVD may be complex (this topic will be discussed in Chapter 2). Excessively increasing OVD beyond physiologically acceptable limits may lead to esthetic and phonetic problems, irritation of the denture foundation, generalized patient discomfort, and neuromuscular symptoms [18].

Prosthetic Factors: Existing Dentures

Critical appraisal of existing prostheses may indicate what should be designed into the new complete dentures and what should be avoided. This information also helps to gauge the limitations of such treatment and determine if patient complaints related to existing complete dentures are justified.

The existing dentures should be assessed carefully for comfort, retention, stability, and support. Occlusion, denture tooth arrangement, and any discrepancy between centric occlusion and maximum intercuspal positions should be evaluated critically. Dental midline placement, arrangement of anterior denture teeth, shade, and type of teeth, occlusal plane orientation, border extensions, cameo surface contours, phonetics, and incisor display should be evaluated and noted in detail. The occlusal surfaces, the intaglio surface, and the cameo surface should be examined for evidence of deterioration and previous repair and / or reline.

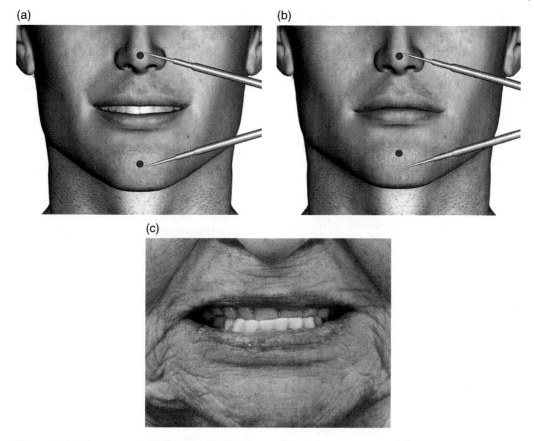

Figure 1.7 (a) Measurement of RVD using calipers to gauge distance between nose and chin in rest position; (b) measurement of OVD using calipers to gauge the distance between nose and chin when teeth are in contact; (c) patient presenting with pseudo class III appearance due to inadequate OVD.

Current dentures should also be examined to assess the patient's capacity and motivation for meticulous denture hygiene [3, 13]. Limitation in this area must be addressed.

Prosthetic Factors: Skeletal Relationship

The anterior-posterior relationship of the maxilla to the mandible should be evaluated in profile to ascertain the relative class I, class II, or class III skeletal relationship [19]. Interarch discrepancies in size and position typically lead to problems establishing adequate occlusion and denture stabilization [19, 20]. For patients with severe Class II, Class III, or transverse skeletal relationships, it is critical to achieve adequate

posterior tooth occlusion to avoid denture instability during empty mouth random occlusal contact [20].

Prosthetic Factors: Saliva

Clinical examination must include assessment for xerostomia (e.g., quality and quantity of saliva, dry lips, shiny and dry intraoral mucosa, angular cheilitis, dorsal fissuring of the tongue). A dental mouth mirror adhering to the tongue or buccal mucosal surfaces during intraoral examination indicates dry mouth. Generally, the patients with significant medical histories including multiple prescribed medications will demonstrate signs of reduced salivary flow [21]. A reduction in salivary flow may be associated with local factors (e.g., salivary gland disorders), systemic factors (e.g., Sjogren's syndrome, AIDS, systemic lupus erythematosis, rheumatoid arthritis, scleroderma, uncontrolled diabetes, thyroid dysfunction, and some neurological disorders), and / or prescription medications [22].

The quantity and quality of saliva affects denture success. Reduced salivary output will interfere with complete denture retention and cause generalized soreness in the denture-bearing soft tissues due to frictional irritation [2, 22, 23]. The quality of saliva should also be considered. Saliva that is ropy, viscous, and mucinous has poor cohesive and adhesive properties prohibiting optimal denture retention. Patients should be informed about these conditions, educated regarding implications, and instructed about necessary treatment, to include the use of denture adhesives, saliva substitutes, and implants for improving the prognosis [13].

Prosthetic Factors: Oral Tolerance

To gauge the patient's oral tolerance (i.e., tendency to gag), a large stock impression tray can be inserted in the patient's mouth. The tray should be inserted gradually, keeping contact / pressure on the ridge crest until the posterior aspect of the tray contacts the soft palate. Reaction of the patient should be observed closely. Slight reaction is normal, but hypersensitivity is a concern. Areas typically related to reflexive gagging include posterior palate, base of tongue, and posterior-lateral tongue borders. Various distraction techniques have developed over the years to overcome the tendency to gag, including asking the patient to lift one of their legs off the chair, or having them chew on ice immediately prior to dental procedures. Prescription of antianxiety medications or application of topical anesthetic (sprays, lozenges, and lollipops) to the soft palate and tongue has been recommended and may be useful in patients with extreme gag reflex.

Prosthetic Factors: Temporomandibular Joints

The temporomandibular joints (TMJs) must be carefully assessed via patient history, joint auscultation, digital palpation, and manual load testing. Radiographs may also be used to investigate symptomatic TMJs. Often patient accommodation and joint adaptation permit relatively normal and pain-free function of TMJs with clinically discernable clicking, popping, and crepitus. The presence of prolonged and debilitating symptoms necessitates further evaluation and referral to a practitioner specializing in the diagnosis and management of temporomandibular dysfunction (TMD).

TMD can be associated with instability of occlusal relationships, oro-facial pain, and functional discomfort. It may also be associated with decrease oral opening (i.e., <60 mm from maxillary anterior ridge to mandibular anterior ridge in edentulous patients is considered to be reduced opening) and deviation of the mandible upon opening. It has been proposed that openings of less than 35–40 mm warrent further investigation [24].

Patients experiencing TMD and TMJ pain are less likely to adapt favorably to new complete dentures. It is therefore important to resolve TMJ pain before initiating definitive complete denture therapy [25]. The TMJ position, range of motion, and function should be appropriate, comfortable, and stable before definitive dental rehabilitation is attempted [25].

Prosthetic Factors: Oral Cancer Review

The lips, cheeks, lateral / ventral tongue surfaces, floor of the mouth, tonsils, soft palate, oropharynx, and neck must be carefully examined and palpated for suspected lesions. Lymph nodes draining the head and neck are of particular interest. Suspected lesions should be digitally palpated to identify bumps, roughness, irregularities, and induration. Not all lesions need to be biopsied, but suspicious findings must be regularly monitored by a dental professional, knowledgeable in pathologic disease progression. Oral cancer screenings should be accomplished periodically for all patients, not just new patients.

Oral Characteristics

Oral Characteristics: Palatal Throat Form

To examine the character, location, and extent of the tissue contour at the junction of the hard and soft palate (i.e., palatal throat form) the patient is asked to open the mouth widely so that this critical palatal area can be observed at relative physiological rest. In order to appreciate the importance of this clinical determination, two concepts must be understood. The first, postpalatal seal area (PPS area), is the soft tissue area at or beyond the junction of the hard and soft palate on which pressure, within physiologic limits, can be applied by a complete denture to aid in denture retention [26]. The second concept, postpalatal seal (PPS), is the region along the posterior border of a maxillary complete denture specifically contoured to facilitate peripheral seal of the prosthesis [26].

Palatal throat form affects peripheral seal of the maxillary complete denture along its posterior border or PPS [3]. Patients possessing a broad band of relatively immovable tissues at junction of the hard and soft palate (Class I) (Figure 1.8a) present a good opportunity to develop a sound PPS facilitating excellent retention of the maxillary denture [3]. Those having a narrower band of immobile tissue and more significant soft palate drape (Class II) (Figure 1.8b) have reduced surface area upon which to develop an effective PPS [3]. Patients that demonstrate severe soft palate drape, even at physiologic rest (Class III) (Figure 1.8c), have minimal surface area at the hard and soft palate junction available for an effective PPS, thus jeopardizing peripheral seal and maxillary denture retention [3]. For individuals with Class III palatal throat form,

precise positioning and careful development of the PPS is critical to achieving and maintaining predictable maxillary denture retention.

Oral Characteristics: Arch Size

Arch size can be measured intraorally or on a cast. A ruler or a Boley gauge can be used to measure the crest-to-crest width and the anterior-posterior length of the edentulous ridges. Large arches (>45 mm width and >55 mm anterior–posterior) offer the potential for optimal retention and stability of complete dentures. Medium-sized arches (approximately 40 mm width and 50 mm anterior-posterior) provide good, but not ideal, characteristics for denture retention and stability. Small arches (<35 mm width and <45 mm anterior-posterior) do not lead to predictable denture retention and stability. A small edentulous arch where the teeth must be positioned facial to the ridge for optimal esthetics and soft tissue support, presents a substantial challenge to prosthesis retention and stability. Arch-size measurements also aid in impression tray selection.

Oral Characteristics: Maxillary Ridge Height

To assess maxillary residual ridge height, a measurement is made from the depth of the labial vestibule to the crest of the edentulous ridge at the midline with the lip gently retracted (Figure 1.9) [4, 5, 27–29]. The vertical ridge dimension undergoes a continuous resorption once the teeth are lost. The amount and rate of ridge resorption in the anterior maxilla depends in large part on the presence of teeth in the mandible. If only anterior mandibular teeth are present, resorption in the anterior maxilla may be significant due to increased and continuous loading forces [30]. This is believed to be the case for patients affected by combination syndrome [31].

Reduced residual ridge height adversely impacts the potential for maxillary denture

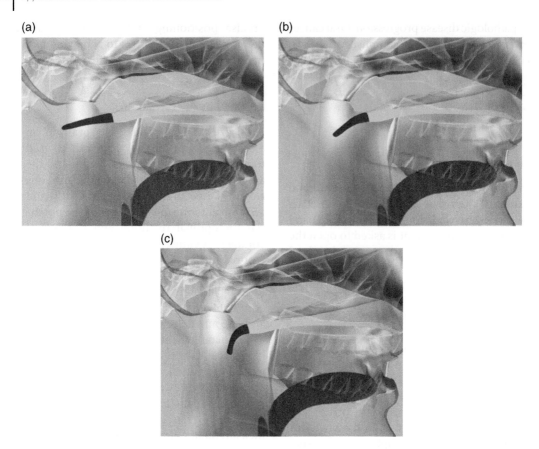

Figure 1.8 (a) Broad palatal throat form, Class I; (b) class II palatal throat form; (c) class III palatal throat form.

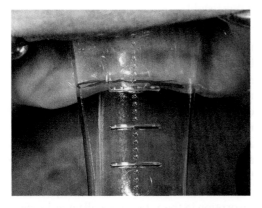

Figure 1.9 Maxillary anterior ridge height measured from depth of labial vestibule to crest of edentulous ridge.

retention and stability [5, 32], which, in turn, has a detrimental effect on muscle tone and esthetics of the patient. Complete denture prognosis could be affected by anterior and posterior maxillary ridge height.

Oral Characteristics: The Palate

Tooth loss and alveolar resorption may lead to alteration of the depth and contour of the palatal vault. Depth and cross-sectional contour of the palatal vault can be evaluated on a dental cast, by intraoral observation, or through intraoral photographs. A flexible transparent ruler is used to record the distance between the deepest aspect of the palate and the most reduced aspect of the ridge crest (Figure 1.10). Broad (i.e., U-shaped) palatal vaults are ideal, offering the potential for excellent support and stability of the maxillary complete denture.

Tapered (i.e., V-shaped) palatal vaults provide less denture stability and are associated with increased processing distortion (i.e., increased denture tooth movement and reduced palatal contact). Flat palatal form provides adequate vertical denture support, but contributes minimally to complete denture stability [20]. Maxillary complete dentures made to fit flat palatal form, particularly when accommodating large labial frenula, render the prosthesis susceptible to fracture. The presence of tori may complicate prosthesis structural integrity and retention. Conventional or zygomatic implants may be necessary to provide adequate prosthesis support, stability, and retention for these patients.

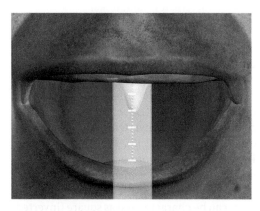

Figure 1.10 Use of a transparent ruler to measure depth of palatal vault.

Oral Characteristics: Maxillary Ridge Contour

Ridge resorption or surgical intervention can affect the cross-sectional form of the residual ridge. Residual ridge resorption (RRR) progressively alters ridge form and size from relatively U-shaped (Figure 1.11a) to knife edged (Figure 1.11b). Further RRR may lead to flat ridges, eventually resulting in depressed or negative ridge form [5]. Maxillary ridge cross-sectional form can be characterized as U-shaped, V-shaped (tapered), round (bulbous), flat, depressed (negative), or any combination of these forms [3, 20, 33]. The shape and contour of the ridge affects retention and stability of the complete dentures.

U-shaped ridges with medium to tall parallel walls and broad, flat ridge crests provide excellent denture retention and stability. U-shaped ridges with short parallel walls and flat ridge crests provide less stability. V-shaped ridges with thin crests or extremely short to flat ridges are typically associated with relatively poor denture support, stability, and retention [33].

Oral Characteristics: The Maxillary Denture Foundation

Oral examination, digital palpation, diagnostic casts, and intraoral photographs (occlusal views) are used to evaluate characteristics of

(a)

(b)

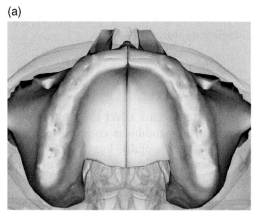
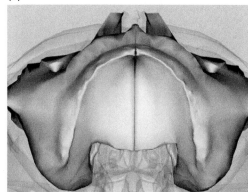

Figure 1.11 (a) U shaped maxillary residual ridge; (b) knife edged maxillary residual ridge.

the denture foundation. The presence of depressed irregularities, exostoses, palatal tori, hypertrophic tuberosities, and significant undercut areas in the maxillary denture foundation should be noted. Surgical intervention should be considered for defects expected to cause chronic soft tissue irritation, restrict normal function, prohibit optimal impression making, or interfere with proper denture border extensions [31, 34].

Oral Characteristics: Mandibular Ridge Height

To assess mandibular residual ridge height, a measurement is made from the depth of the labial vestibule to the crest of the edentulous ridge at the midline with the lip gently retracted (Figure 1.12) [4, 5, 27–29]. It is important to avoid distending the vestibule while recording this measurement. The amount of force imparted to the mandible during normal functional loading may be twice that for the maxilla due to the reduced surface area of the denture foundation. This, in part, is believed to account for the increased RRR experienced by the mandible as compared to the maxilla [5, 29].

Loss of residual ridge height adversely affects the complete denture retention and stability [5]. For some patients, mandibular RRR can be so extreme that the mandible is susceptible to pathologic fracture. Advanced RRR complicates both the dentist's ability to

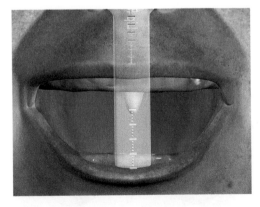

Figure 1.12 Measurement of height of mandibular anterior ridge using transparent ruler.

fabricate adequate complete dentures and the patient's ability to manage new prostheses successfully. It is therefore always prudent to consider means of maintaining and improving the denture foundation by the retention of natural tooth roots for conventional complete overdentures to slow the rate of RRR, or strategically place dental implants to improve the mechanics of denture support, retention, and stability.

Complete denture prognosis on the basis of radiographic mandibular bone height may be determined as follows: ≥21 mm (Class I) favorable, 16–20 mm (Class II) acceptable, 11–15 mm (Class III) compromised, and ≤10 mm (Class IV) guarded [4].

Oral Characteristics: Mandibular Ridge Contour

Ridge resorption or surgical intervention can affect the cross-sectional form of the mandibular residual ridge. RRR progressively alters ridge form and size from relatively robust (inverted U-shape) (Figure 1.13a), to significantly diminished (inverted V-shape), to knife edged (Figure 1.13b) [5]. Further mandibular RRR resorption may produce flat (Figure 1.13c) or even depressed (negative) ridge form [5].

Therefore, mandibular ridge cross-sectional form can be characterized as square (inverted U-shape), tapered (inverted V-shape), round (bulbous), flat, depressed (negative), or any combination of these forms [3, 20, 33]. The shape and contour of the ridge affects expected retention and stability of the complete dentures.

Oral Characteristics: Mandibular Muscle Attachments

Muscle attachments affect the contour and extension of mandibular complete denture flanges [2]. Unfavorable location of muscle attachments will have a detrimental effect on denture stability. In such circumstances, surgical correction should be considered [14, 35]. The amount of RRR alters the relative relationship of the muscle attachments

(a)

(b)

(c)

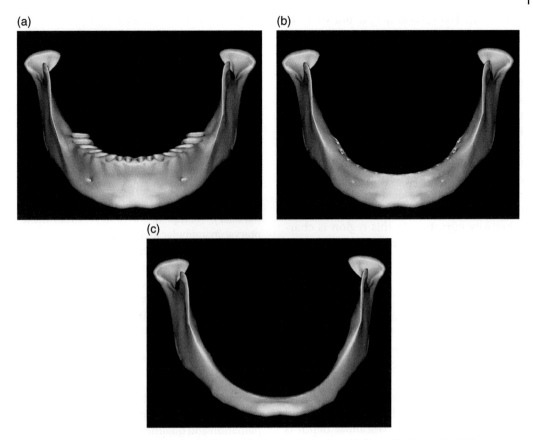

Figure 1.13 (a) Inverted U shaped mandibular residual ridge; (b) knife edged mandibular residual ridge; (c) flat mandibular residual ridge.

to the residual ridge crest. It is important to appreciate that this relationship changes with ongoing ridge resorption [3]. Mandibular muscle attachments are therefore classified as low (near the vestibular reflection), middle, or high (near the ridge crest) [3].

Oral Characteristics: Mandibular Denture Foundation

The existence of depressed irregularities, exostoses, lingual tori, and significant undercut areas within the mandibular denture foundation should be noted. Surgical correction is considered for defects expected to cause chronic soft tissue irritation, restrict with normal function, prohibit optimal impression making, or interfere with proper denture border extensions [31, 34].

In extreme cases, mandibular resorption is so advanced that the ridge appears flat, or even concave, and palpation of the mylohyoid ridge reveals a sharp spinelike projection with thin soft tissue covering. This area is highly susceptible to functional stress imparted by the denture, and surgical correction must be considered [35].

Oral Characteristics: Maxillary Tuberosity Curve

Contours of the maxillary tuberosities, from ridge crest to vestibular depth, are assessed by digital palpation or dental mirror placement into the distal extent of the maxillary vestibule and instructing the patient to open their mouth and move their jaw from side to side. This space has been called the retrozygomatic space or the corono-maxillary space [36]. The vertical height and width of this space varies with mouth opening and must be carefully considered when molding borders and making impressions [36]. Definitive denture borders in this area should account for the dynamic nature of this space during mandibular movements. Failure to do so will result in an inadequate peripheral seal. Excessive flange thickness in this area will result in discomfort and/or denture displacement as the coronoid process impinges on the denture flange during lateral mandibular movements. Maxillary tuberosity curvature in this region is characterized as flat, moderately curved, steep, or undercut.

Oral Characteristics: Vestibule

Both RRR and the location of the muscle attachments affect relative vestibular depth [2, 5]. Unfavorable (shallow) vestibular depth has a detrimental effect on the complete denture stability and due consideration should be given to corrective preprosthetic surgery (i.e., vestibuloplasty) [35] or dental implant placement. The vestibular depth is classified as deep, average, or short.

Oral Characteristics: Frenula Attachments

Frenula attachments affect the shape and the extension of the complete denture flanges [2]. Frenula attached near the edentulous ridge crest can be a focus of irritation if not accommodated by flange contour [35]. Overrelief of the flange during denture adjustment may lead to the ingress of air, loss of peripheral seal, and compromised denture retention [35]. Excessive flange notching to accommodate frenula attached near the ridge crest can concentrate loading stress, resulting in premature denture-base failure [35]. Therefore, corrective preposthetic surgery for frenulum attachments near the ridge crest should be considered [14]. Maxillary frenulum/muscle attachments are therefore classified as high (near the vestibular reflection), middle, and low (near the ridge crest).

Oral Characteristics: Pterygomandibular Raphe

The pterygomandibular raphe is a vertically positioned tendinous band coursing bilaterally from the hamuli of the medial pterygoid plates to the posterior limit of the mandibular retromolar trigones. It serves as a facial raphe between a portion of the buccinator muscle and the ipsilateral superior pharyngeal constrictor. The nature of attachment of the pterygomandibular raphe within the pterygomaxillary (hamular) notch will affect the shape and the extension of the posterior-lateral aspects of the maxillary complete denture. Unfavorable pterygomandibular raphe attachment (i.e., near the ridge/tuberosity crest) will have a detrimental effect on the stability and retention of the denture. Attachment of the pterygomandibular raphe is therefore classified as high (deep in the pterygomaxillar notch), average, or low (near the ridge/tuberosity crest).

Oral Characteristics: Denture Bearing Soft Tissues

Compressibility of the soft tissues of the denture foundation can be assessed by digital palpation and qualified as severely compressible, moderately compressible, slightly compressible, or thin and delicate (Figure 1.14a).

(a)

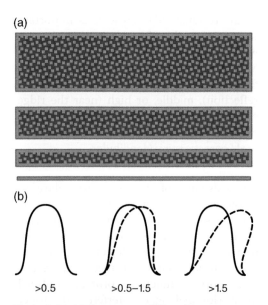

(b)

>0.5   >0.5–1.5   >1.5

Figure 1.14 (a) The compressibility of denture-bearing soft tissues based on thickness. (b) Characterization of soft tissue displaceability.

(a) (b)

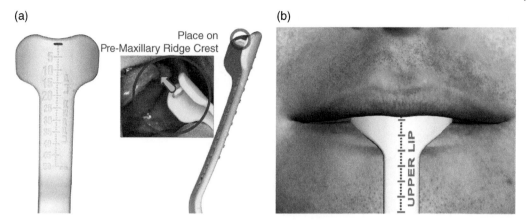

Figure 1.15 (a) Lip ruler; (b) Measurement of maxillary esthetic space made with a lip ruler at rest.

Mobility or soft tissue displacement may be characterized as severely displaceable (>1.5 mm), moderately displaceable (between 0.5–1.5 mm), or slightly displaceable (<0.5 mm) (Figure 1.14b). Mobilizing the tissue using two mouth mirror handles permits assessment of soft tissue displacement. Clinical soft tissue thickness may be qualified as severely compressible and easily displaceable (thick and spongy), moderately compressible and moderately displaceable (2–3 mm thick), or noncompressible and nondisplaceable (relatively thin). Soft tissues that are noncompressible and nondisplaceable offer little denture support, are highly susceptible to irritation under pressure, and compromise denture retention [10]. Severely compressible and easily displaceable tissues are associated with excessive denture movement and should be considered for surgical correction [27].

Oral Characteristics: Retromolar Pads

Compressibility of the retromolar pads can be assessed through digital palpation or exploration with a blunt instrument, as severely compressible, moderately compressible, slightly compressible, or thin and delicate. Lateral mobility or displacement of retromolar pads can be classified in similar fashion as severely displaceable (>1.5 mm),

moderately displaceable (between 0.5–1.5 mm), or slightly displaceable (<0.5 mm).

Oral Characteristics: Maxillary Ridge Crest to Resting Lip Length (Esthetic Space)

A lip ruler (Figure 1.15a) is used to measure the distance between the maxillary edentulous ridge crest at the midline and upper lip at rest (Figure 1.15b) and during smile [37]. This measurement provides information on the potential for and extent of maxillary anterior denture tooth display and provides information to laboratory technician regarding wax rim length and denture tooth setup [37].

Oral Characteristics: Mandibular Ridge Crest to Resting Lip Length (Esthetic Space)

A lip ruler is used to measure the distance between the mandibular edentulous ridge crest at the midline and the lower lip at rest and during smile [37]. This measurement permits the practitioner to gauge the display of the mandibular anterior denture teeth, and provides important information for the laboratory technician [37].

Oral Characteristics: Maximal Oral Opening

A triangular measuring gauge is used to measure the interarch distance at the midline during maximum oral opening (Figure 1.16a,

(a) (b)

(c)

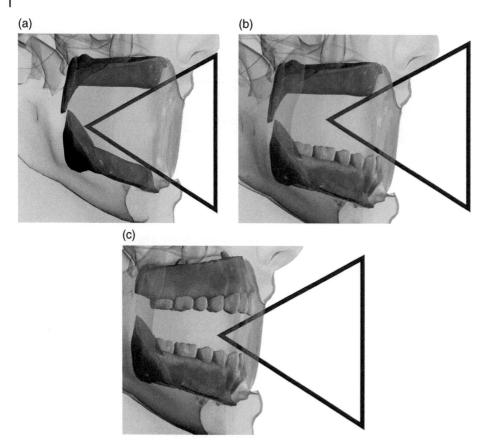

Figure 1.16 (a) Triangular shaped measuring gauge used to measure interarch (ridge to ridge) distance during maximum oral opening; (b) triangular shaped measuring gauge used to measure interarch (ridge to tooth) distance during maximum oral opening; (c) triangular shaped measuring gauge used to measure interarch (tooth to tooth) distance during maximum oral opening.

b and c). Maximum interarch (ridge-to-ridge) distance of ≥60 mm is considered acceptable for edentulous patients. An interarch (ridge-to-tooth) distance of ≥50 mm is considered acceptable for patients edentulous in one arch. An interarch (tooth-to-tooth) distance ≥45 mm is considered acceptable for dentate patients [38].

Oral Characteristics: Retromylohyoid Space

The retromylohyoid space, commonly referred to as lateral throat form, is a bilateral potential space immediately lingual to the retromolar pads bounded anteriorly by the mylohyoid ridge and muscle, posteriorly by the retromylohyoid curtain, inferiorly by the floor of the lingual vestibule, and lingually by the anterior tonsillary pillar, when the tongue is relaxed [39, 40]. The degree to which this potential oral space can be occupied by posterior extension of the mandibular complete denture lingual flange will influence mandibular denture stability [39].

Lateral throat form is evaluated by placing a dental mirror into the retromyloyoid space (Figure 1.17), instructing the patient to project the tongue tip to the contralateral oral commissure, and observing the degree of mirror displacement. Lateral throat form is qualified as deep (no mirror displacement), medium (minor mirror displacement), or shallow (maximal mirror displacement). Extension of the mandibular denture lingual flange deep into the lateral throat form contributes favorably to denture stability [39, 40].

Oral Characteristics: Tongue Size

To examine tongue size and characterize it as extra large, large, average, or small, the patient is instructed to open the mouth as if to receive food [41–43]. Patients who have been edentulous for an extended time tend to develop a flat and broad (large) tongue [2]. Placement of a mandibular complete denture in such patients results in complaints of crowding, discomfort, and inadequate tongue space [44]. Initially, these patients find it

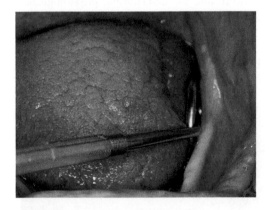

Figure 1.17 Head of the mirror placed in retromylohyoid space for assessing lateral throat form.

difficult to adjust to the new mandibular denture and constantly dislodge the denture through uncoordinated tongue movements. Fortunately, with the passage of time and experience in denture wearing, patient and tongue adaptations permit relatively successful mandibular denture stability and function.

Oral Characteristics: Tongue Position

To observe the natural tongue position, the patient is instructed to open the mouth as if to receive food [41–43]. Care should be taken to avoid mention of the word "tongue" so as not to draw the patient's attention to the purpose of the examination [41–43]. Observation of tongue position will permit qualification as normal or retracted. Normal position is demonstrated when the tongue completely fills the floor of the mouth, the lateral borders rest over the posterior edentulous ridges, and the tip of the tongue rests on or just lingual to the anterior mandibular ridge crest (Figure 1.18a). Retracted posture is indicated when the tongue is pulled back into the mouth exposing the floor of the mouth and lateral tongue borders lie medial or posterior to the edentulous ridge (Figure 1.18b). In addition, the tip of the tongue in retracted posture is either located in the posterior aspect of the oral cavity or withdrawn into the body of the tongue. Approximately two-thirds of patients present with normal

(a)

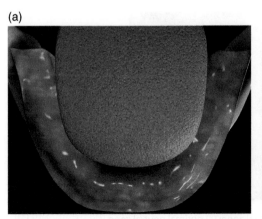

(b)

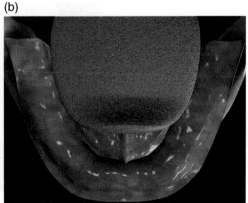

Figure 1.18 (a) Normal tongue position; (b) retracted tongue position.

tongue posture and one-third with retracted tongues [41–43].

Tongue position influences mandibular complete denture flange design and general denture stability [41–43]. Normal tongue position favorably postures the floor of the mouth for predictable lingual flange extension and contour, permitting maintenance of peripheral denture seal, and increasing denture stability and retention. Denture prognosis for patients with retracted tongues may be improved by making the patient aware of this condition and instructing them to consciously maintain normal tongue posture for improved denture retention and stability. The use of tongue exercises and training contours have been suggested to aid in improved tongue posture [41–43, 45].

Oral Characteristics: The Neutral Zone

The neutral zone is an area within the oral cavity where outward forces originating from the tongue are neutralized by inward forces originating from the lips and cheeks [46–49]. The approximate facio-lingual width of the neutral zone can be evaluated by instructing the patient to open the mouth, permitting assessment of available space between the tongue and adjacent lips / cheeks. In so doing, the neutral zone may be qualified as restricted (Figure 1.19a), reduced (Figure 1.19b), or optimal (Figure 1.19c). As will be detailed later in this text, the neutral zone is used as a convenient guide to develop physiologic contours for the polished surfaces of the mandibular denture and for determining physiologically appropriate facio-lingual tooth positions.

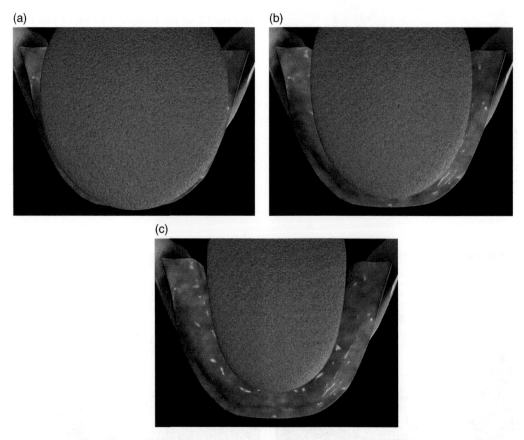

Figure 1.19 (a) Restricted neutral zone; (b) reduced neutral zone; (c) optimal neutral zone.

Summary

Specific factors discernible during careful and detailed examination of edentulous patients permit development of an accurate therapeutic prognosis and provide critical information for optimal treatment. Following thorough patient assessment, the treatment prognosis should be classified as optimal, moderate, compromised, or guarded. Improvement of the denture foundation by tissue conditioning, preprosthetic surgery, and / or placement of dental implants may enhance the prognosis. Healthy and stable temporomandibular joints improve the overall prognosis and are a prerequisite for definitive prosthodontics therapy.

The assumption that patients understand our diagnostic findings and treatment recommendations can be a major cause of patient dissatisfaction. Patients must be thoroughly educated and regularly reminded with respect to compromising factors identified during the initial assessment that will adversely affect treatment and expected outcomes. Patients informed early in the therapeutic process appreciate obstacles to optimal treatment, whereas explanations provided only after problems are encountered tend to be looked upon as excuses. It is important to avoid initiating therapy for patients who do not understand, or refuses to accept, limitations of proposed treatment. Additionally, patients must appreciate fees, prosthesis replacement frequency, and regular maintenance requirements as a critical element of informed consent prior to initiating treatment. To this end, a software application incorporating a carefully organized examination form has been developed by the authors to aid in examination, diagnosis, treatment planning, and prognosis for complete denture patients.

References

1 Sato, Y., Tsuga, K., Akagawa, Y., and Tenma, H. (1998) A method for quantifying complete denture quality. *J Prosthet Dent*, **80**, 52–57.

2 Barone, J. V. (1964) Diagnosis and prognosis in complete denture prosthesis. *J Prosthet Dent*, **14**, 207–213.

3 House, M. M. (1958) The relationship of oral examination to dental diagnosis. *J Prosthet Dent*, **8**, 208–219.

4 McGarry, T. J., Nimmo, A., Skiba, J. F. *et al.* (1999) Classification system for complete edentulism. *J Prosthodont*, **8**, 27–39.

5 Atwood, D. A. (1971) Reduction of residual ridges: A major oral disease entity. *J Prosthet Dent*, **26**, 266–279.

6 Neil, E. (1932) *Full Denture Practice.* Marshall & Bruce, Nashville, TN, pp. 14–16, 81–83.

7 Winkler, S. (2005) House mental classification system of denture patients: The contribution of Milus M. *House. J Oral Implantol*, **31**, 301–303.

8 Koper, A. (1970) The initial interview with complete-denture patients: its structure and strategy. *J Prosthet Dent*, **23**(6), 590–597.

9 Engelmeier, R. L., and Phoenix, R. D. (1996) Patient evaluation and treatment planning for complete-denture therapy. *Dent Clin North Am*, **40**, 1–18.

10 Jamieson, C. H. (1958) Geriatrics and the denture patient. *J Prosthet Dent*, **1**, 8–13.

11 Pickett, H. G., Appleby, R. G., and Osborn, M. O. (1972) Changes in the denture supporting tissues associated with the aging process. *J Prosthet Dent*, **27**(3), 257–262.

12 Silverman, S. (1958) Geriatrics and tissue changes – problem of the aging denture patient. *J Prosthet Dent*, **8**, 734–739.

13 Lang, B. R. (1994) A review of traditional therapies in complete dentures. *J Prosthet Dent*, **72**, 538–542.

14 Niiranen, J. V. (1954) Diagnosis for complete dentures. *J Prosthet Dent*, **4**, 726–738.

15 Ettinger, R. L. (2014) Dentures, in *The Encyclopedia of Elder Care. The Comprehensive Resource on Geriatric Health and Social Care*, 3rd edn (eds.: E. A. Capezutti, M. L. Malone, P. R. Katz, and M. D. Mezey). Springer, New York, NY, pp. 211–214.

16 Gibson, R. M. E. (1967) Panographic survey of edentulous mouths. Thesis. University of Michigan, Ann Arbor, MI, p. 43

17 Niswonger, M. E. (1934) The rest position of the mandible and the centric relation. *J Am Dent Assoc*, **21**, 1572–1582.

18 Massad, J. J., Connelly, M. E., Rudd, K. D., *et al.* (2004) Occlusal device for diagnostic evaluation of maxillomandibular relationships in edentulous patients: a clinical technique. *J Prosthet Dent*, **91**, 586–590.

19 Angle, E. H. (1899) Classification of malocclusion. *Dent Cosmos*, **41**, 248–264.

20 Jacobson, T. E., and Krol, A. J. (1983) A contemporary review of the factors involved in complete dentures. Part II: Stability. *J Prosthet Dent*, **49**, 165–172.

21 Wiener, R. C., Wu, B., Crout, R., *et al.* (2010) Hyposalivation and xerostomia in dentate older adults. *J Am Dent Assoc*, **141**, 279–284.

22 Sreebny, L. M. (1968) The role of saliva in prosthodontics. *Int Dent J*, **18**, 812–822.

23 Kawazoe, Y., and Hamada, T. (1978) The role of saliva in retention of maxillary complete dentures. *J Prosthet Dent*, **40**, 131–136.

24 Sheppard, I. M., and Sheppard, S. M. (1965) Maximal incisal opening – a diagnostic index? *J Dent Med*, **20**, 13–15.

25 McHorris, W. H. (1974) TMJ dysfunction – Resolution before reconstruction. *J Eur Acad Gnathology*, **1**, 16–32.

26 The glossary of prosthodontic terms, 8th edition (2005) *J Prosthet Dent*, **94**, 63.

27 Carlsson, G. E. (1998) Clinical morbidity and sequelae of treatment with complete dentures. *J Prosthet Dent*, **79**, 17–23.

28 Costello, B. J., Betts, N. J., Barber, H. D., and Fonseca, R. J. (1996) Preprosthetic surgery for the edentulous patient. *Dent Clin North Am*, **40**, 19–38.

29 Tallgren, A. (1972) The continuing reduction of the residual alveolar ridges in complete denture wearers: a mixed-longitudinal study covering 25 years. *J Prosthet Dent*, **27**(2), 120–132

30 Smith, D. E., Kydd, W. L., Wykhuis, W. A., and Phillips, L. A. (1963) The mobility of artificial dentures during comminution. *J Prosthet Dent*, **13**, 839–856

31 Kelly, E. K. (1966) The prosthodontist, the oral surgeon, and the denture-supporting tissues. *J Prosthet Dent*, **16**, 464–478.

32 Jahangiri, L., Devlin, H., Ting, K., and Nishimura, I. (1998) Current perspectives in residual ridge remodeling and its clinical implications: a review. *J Prosthet Dent*, **80**(2), 224–237.

33 Appleby, R. C., and Ludwig, T. F. (1970) Patient evaluation for complete denture therapy. *J Prosthet Dent*, **24**, 11–17.

34 Yrastorza, J. A. (1963) Surgical problems in edentulous jaws associated with denture construction: a review. *J Oral Surg Anesth Hosp Dent Serv*, **21**, 202–209.

35 Miller, E. L. (1976) Sometimes overlooked: preprosthetic surgery. *J Prosthet Dent*, **36**, 484–490.

36 Arbree, N. S., Yurkstas, A. A., and Kronman, J. H. (1987) The coronomaxillary space: literature review and anatomic description. *J Prosthet Dent*, **57**(2), 186–190.

37 Massad, J. J., Ahuja, S., and Cagna, D. (2013) Implant overdentures: selections for attachment systems. *Dent Today*, **32**(2), **128**, 130–132.

38 Sheppard, I. M., Sheppard, S. M. (1965) Maximal incisal opening – a diagnostic index? *J Dent Med*, **20**, 13–15.

39 Levin, B. (1981) Current concepts of lingual flange design. *J Prosthet Dent*, **45**, 242–252.

40 Neil, E. (1941) *The Upper and the Lower – A Simplified Full Denture Impression Procedure*. Chicago: Coe Laboratories, Inc., pp. 75–149.

41 Wright, C. R., Swartz, W. H., and Godwin, W. C. (1961) *Mandibular Denture Stability – A New Concept.* Ann Arbor: The Overbeck Co. Publishers, pp. 7–17.

42 Wright, C. R. (1966) Evaluation of the factors necessary to develop stability in mandibular dentures. *J Prosthet Dent*, **16**, 414–430.

43 Wright, C. R., Muyskens, J. H., Strong, L. H., *et al.* (1949) A study of the tongue and its relation to denture stability. *J Am Dent Assoc*, **39**, 269–275.

44 Kingery, R. H. (1936) Examination and diagnosis preliminary to full denture construction. *J Am Dent Assoc*, **23**, 1707–1713.

45 Kuebker, W. A. (1984) Denture problems: causes, diagnostic procedures, and clinical treatment. I. Retention problems. *Quintessence Int*, **10**, 1031–1044.

46 Fish, E. W. (1933) Using the muscles to stabilize the full lower denture. *J Am Dent Assoc*, **20**, 2163–2169.

47 Fish, E. W. (1964) *Principles of Full Denture Prosthesis*, 6th edn. Staples, London, pp. 36–37.

48 Schiesser, F. J. (1964) The neutral zone and polished surfaces in complete dentures. *J Prosthet Dent*, **14**, 854–865.

49 Ries, G. E. (1978) The neutral zone as applied to oral and facial deformities, in *Neutral Zone in Complete and Partial Dentures* (eds V. E. Beresin and F. J. Schiesser). 2nd edn. Mosby, St. Louis, MO, pp. 208–220.

2

Orthopedic Resolution of Mandibular Posture

Introduction

Gradual and continuous reduction of occlusal vertical dimension (OVD) is commonly seen with extended use of complete dentures [1]. Residual ridge resorption (RRR) and wear of prosthetic teeth are the two major causative factors for loss of OVD in complete denture wearers [2–4]. Patients demonstrating insufficient OVD may present with the following clinical signs and symptoms: reduction of lower face height, compromised facial esthetics and function, and acquired Class III maxillo-mandibular relationship [5], angular cheilitis and / or temporo-mandibular joint (TMJ) sounds on auscultation [1]. One therapeutic objective for these patients is to re-establish an esthetically appropriate OVD [1]. Potential problems associated with increasing the OVD during fabrication of new complete dentures may include altered phonetic and masticatory function, unacceptable facial appearance, muscular discomfort, accelerated alveolar bone loss, sore residual ridges, premature and excessively audible denture tooth contact, and exaggerated gagging [6–14].

Habitual mandibular posturing and neuromuscular programming in patients with insufficient OVD may also present challenges in registering the optimal maxillo-mandibular relationship [1]. Attempts at registering the OVD while fabricating new dentures may lead to recording an inadequate OVD due to the present condition or recording an excessively increased OVD in an attempt to reestablish esthetics [1]. If the OVD is increased beyond the physiologic tolerance of the patient in the new set of complete dentures it can lead to several problems. Patients may develop altered phonetics, muscle tenderness, exaggerated gagging, premature contact and clicking of denture teeth, bulky sensation, soreness, and increased bone loss [6–14].

For patients demonstrating insufficient OVD in their existing prostheses, it is important to determine and test the most appropriate interarch jaw relationship during the diagnostic phase of therapy. This can be accomplished by fabricating a treatment denture or by modification of the existing denture [15–20]. Alteration of the existing prosthesis is more economical and less time consuming than fabricating a new treatment prosthesis. The altered prosthesis may be called an orthopedic occlusal device as it maintains the orthopedic position of the mandible. The re-established position permits evaluation of the proposed maxillo-mandibular relationship and facilitates deprogramming of the muscles, allowing the patient to adjust to the altered interarch relationship prior to delivering the new prosthesis. The dental practitioner and the patient can also assess esthetics, function, and patient comfort at the proposed OVD. Only when conditions are judged appropriate by the clinician and acceptable to the patient should the fabrication of new complete dentures be initiated [21–24].

Application of the Neutral Zone in Prosthodontics, First Edition. Joseph J. Massad, David R. Cagna, Charles J. Goodacre, Russell A. Wicks and Swati A. Ahuja.
© 2017 John Wiley & Sons, Inc. Published 2017 by John Wiley & Sons, Inc.
Companion website: www.wiley.com/go/massad/neutral

Fit of the existing complete dentures and the health of the intraoral tissues should be optimized prior to conversion of the denture to an orthopedic occlusal device [25]. The soft tissues supporting ill-fitting complete dentures are often displaceable, irritated, easily traumatized, and / or grossly deformed [26]. Definitive impressions and interarch relationship records made in the presence of suboptimal supporting tissues are most likely inaccurate. Appropriate application of conditioning materials can improve the health of the denture foundation and optimize fit of the prostheses [26–28].

Conditioning Abused Tissues and Stabilizing the Existing Prosthesis

Soreness, inflammation and / or irritation of the soft tissues supporting the denture foundation is often associated with abuse from an ill-fitting prosthesis (Figure 2.1) [26]. Pressure from irregular intaglio surface topography, occlusal disharmony, avitaminosis, or debilitating diseases can all contribute to soreness and inflammation of the soft tissues. This condition is commonly exacerbated by poor oral hygiene, smoking, and excessive presence of microorganisms.

Microbial colonization and formation of biofilms on acrylic denture base resin materials in the oral environment leads to denture

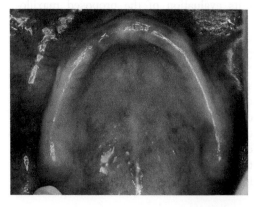

Figure 2.1 Tissues abused from current denture.

stomatitis. Although there is evidence that several specific pathogens are elemental in denture plaque formation, the classic opportunist organism, *Candida albicans*, is most often associated with denture stomatitis *in vivo*. The health of the denture supporting tissues must be optimized prior to the fabrication of new dentures [28]. Patients should be instructed to massage the edentulous ridges regularly with a rotary toothbrush to help abused tissues to optimally recover, clean the dentures twice daily and leave the dentures out of the mouth for 48–72 hours prior to the clinical appointments [28].

Tissue-conditioning materials aid in improving the health of the denture-bearing tissues, act as a cushion between the denture base and the denture bearing mucosa, and improve the fit of the existing denture during the course of fabrication of the definitive prostheses. A perceived disadvantage of these materials is the need to replace them frequently; therefore, the authors suggest the use of ethyl methacrylate homopolymer (PermaSoft Denture Liner, Dentsply Intl) for conditioning the tissues, as this material requires only annual replacement.

Materials Properties and Technique

Tissue-conditioning materials such as ethyl methacrylate homopolymer are most commonly available as a powder / liquid system. The viscoelastic gel formed by mixing the powder and the liquid has the property of bonding to the polymer resin. They continue to flow and be resilient until a substantial amount of alcohol has been depleted. The intaglio surface of the denture should be accurately and evenly relieved to provide adequate space (approximately 1–2 mm) for the resilient conditioning material to be effective [28]. The procedure describing the use of tissue conditioning material is as follows.

Technique

1) Occlusal vertical dimension of the existing prostheses is recorded utilizing a Boley gauge. This OVD is maintained

during the tissue conditioning procedure. Tissue conditioning procedure is performed for its intended purpose – to treat the abused tissues and produce as little change in the OVD as possible.

2) Extensions of the complete denture borders and contours of the dentures base should be assessed for accuracy. Overextended denture borders should be reduced appropriately with laboratory rotary instruments. Small peripheral extension deficits may be compensated by the conditioning materials. Light polymerized resin is ideal for larger corrections.

   a) A suitable bonding agent is applied to the flanges and denture base requiring modification.

   b) Light polymerized resin is adapted to the denture base and the flanges needing modification.

   c) The maxillary and mandibular dentures are best modified one at a time to ensure optimal extension of the flanges of the denture.

   d) Border molding movements (described in Chapter 3 in detail) are performed by the patient to generate physiologically extended denture borders.

   e) Polymerization of resin is activated using a curing light. Care is taken to expose the resin without distorting its position.

   f) The denture is removed from the oral cavity and the resin addition is inspected for adaptation and contour. The denture can be further placed in a polymerization unit to complete the cure of the added light-activated material.

   g) Surface irregularities are adjusted and polished as needed.

3) It is recommended that the intaglio surface be relieved uniformly and adequately to provide adequate space for the tissue conditioning material. Application of a dye solution helps to identify the intaglio surface (Figure 2.2a) and to ensure even and uniform reduction of the intaglio surface and the borders of the complete denture. Standard laboratory rotary instruments can be used to remove denture base material evenly and adequately to optimize fit (Figure 2.2b). A suitable lubricant is applied to all the surfaces of the denture except the intaglio surface and the denture borders. The lubricant will facilitate the removal of the excess tissue conditioning material and finishing of the prostheses.

4) A long-lasting tissue-conditioning material is mixed as per the manufacturer's recommendations and poured into the denture and distributed evenly on the intaglio surface and the borders of the complete denture with a cement spatula.

(a)

(b)

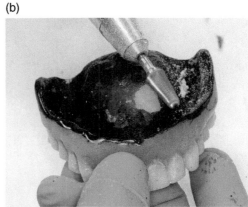

Figure 2.2 (a) Dye solution evenly painted on intaglio surface of maxillary denture; (b) intaglio surface relieved using laboratory rotary instruments.

(It is recommended to perform tissue conditioning for one denture at a time.) Both the dentures are placed in the oral cavity and the patient is guided into the centric relation (CR) position. The denture with the tissue-conditioning material is properly oriented by occluding with the opposing prosthesis. After one prosthesis is lined, then the process is repeated with the opposing denture.

5) Border-molding movements are accomplished again to generate physiologically extended borders.

6) Following the initial polymerization of the tissue conditioning material the denture is removed from the oral cavity. It is then placed in a pressure pot containing warm water for 20 minutes and is pressurized at 20 pounds/inch to diminish the residual monomer.

7) The intaglio surface is examined and excess material is trimmed with sharp scissors or a blade (Figure 2.3).

Tissue-conditioning material should have a uniform thickness of 1.0–1.5 mm. Esthetics, retention and stability of the denture will be compromised if the thickness of the tissue conditioning material exceeds 2.0 mm. If denture base material shows through the conditioning material, the denture base is relieved in those areas with e-cutters and the procedure is repeated until an even layer of

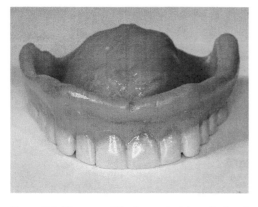

Figure 2.3 Tissue conditioning material applied and functionally adapted on maxillary denture.

tissue conditioned material is available to provide the desired results. Following the tissue-conditioning procedure, the complete denture should accurately contact the denture-bearing tissues with acceptable border extensions to assure optimal retention, stability and comfort.

Re-establishing Orthopedic Mandibular Position

It is critical to establish the appropriate interarch relationship in patients wearing complete dentures (CDs) with insufficient OVD before initiating the process of fabricating the definitive prostheses [1]. An appropriately fabricated orthopedic occlusal device enables examination and evaluation of the patient esthetics, comfort, phonetics and function at the proposed OVD prior to fabrication of new prostheses [1]. It also aids in deprogramming the habitual mandibular posture caused due to the occlusal disharmony in the existing dentures. The procedure for fabricating the orthopedic occlusal device using the intraoral gothic arch tracer is described below:

1) The first step is to verify the need for fabricating the orthopedic occlusal device. To do this we must determine the rest vertical dimension (RVD).

2) Rest vertical dimension can be registered using the exhaustive technique by marking a dot on the tip of the patient's nose and another dot on the prominence of the chin. With the dentures in the patient's mouth, the patient is asked to inhale and exhale several times, until the mandibular muscles fatigue and the mandible droops down [29, 30]. A caliper is used to record the distance between the two dots. This measurement represents the patient's physiological rest position (PRP) or RVD (Figure 2.4a). This procedure can be performed several times to establish a repeatable measurement. The patient is asked to close the mouth (with the existing

(a)

(b)

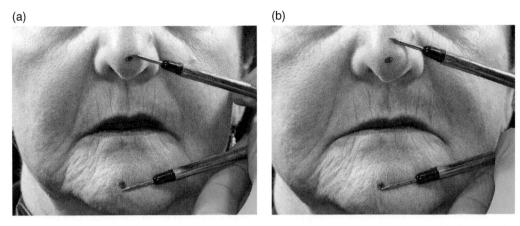

Figure 2.4 (a) Recording of rest vertical dimension; (b) existing occlusal vertical dimension.

Figure 2.5 Striking plate mounted on maxillary denture and pin receiver mounted on mandibular denture.

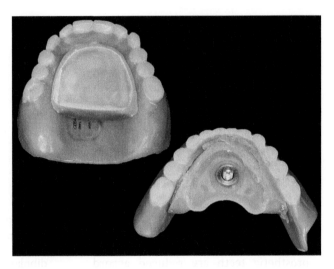

dentures in the mouth) and the distance between the two dots is measured using the calipers. This measurement represents the OVD (Figure 2.4b). The difference between the RVD and the OVD represents the interocclusal distance. Interocclusal distance varies depending on the patient's age, physical and emotional condition, fatigue, and time of the day [31]. As per Niswonger, the OVD should generally be 2.0 to 4.0 mm less than the RVD. An increased interocclusal distance along with the accompanying clinical signs and symptoms corresponds to insufficient OVD and justifies the need for fabrication of an orthopedic occlusal device in the patient. To fabricate the orthopedic occlusal device accurately, the mandible must be vertically located at the desired OVD (2–4 mm less than the RVD) during denture modification. The orthopedic occlusal device is fabricated by the functionally generated path method using the intraoral gothic arch tracer.

3) Following tissue-conditioning, the gothic arch striking plate is attached to the maxillary denture, and the threaded vertical pin receiver with pin is attached to the mandibular denture using light polymerized or auto-polymerizing acrylic resin, as per manufacturer's recommendations (Figure 2.5).

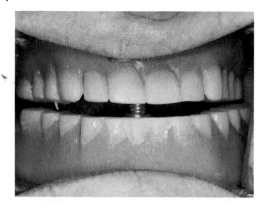

Figure 2.6 Dentures in CR with central bearing pin at the desired OVD.

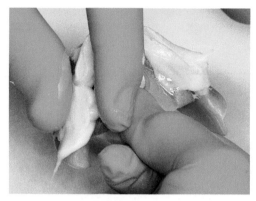

Figure 2.7 Freshly mixed resin placed over the occlusal surfaces of mandibular posterior teeth.

4) The maxillary and mandibular dentures with the attached intraoral tracer assembly are placed in the patient's mouth.

5) The patient is instructed to close slowly until the tip of the threaded pin touches the striking plate. It is crucial to evaluate the OVD at this point. The pin can be moved up or down by rotation to establish the desired OVD and the interocclusal distance (Figure 2.6). The patient is guided to the CR position and the space between the maxillary and the mandibular posterior teeth is analyzed.

6) If this space is less than 2.0–3.0 mm, the occlusal surfaces of the mandibular prosthetic teeth are reduced accordingly with laboratory rotary instruments. Maxillary buccal cusps are adjusted vertically to accentuate the palatal cusps as dominant cusps [32]. If the space is adequate, the occlusal surfaces of the mandibular posterior teeth are roughened with laboratory rotary instruments.

7) A suitable lubricant is added to the occlusal/incisal surfaces of all maxillary teeth. An inking solution is applied to the maxillary striking plate.

8) Monomer is painted on the occlusal surface of the mandibular posterior teeth. Ethyl methacrylate autopolymerizing tooth-colored acrylic resin is mixed and the resin is placed over the occlusal surfaces of man-

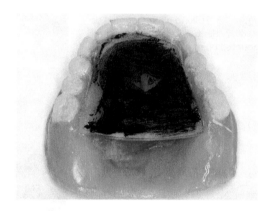

Figure 2.8 Path of mandibular movement visualized on striking plate.

dibular posterior teeth (Figure 2.7). While the acrylic is in a "doughy" consistency, the patient is instructed to close slowly until the tip of the threaded pin touches the striking plate. The patient is asked to perform a full range of mandibular movements (keeping contact between the pin and the striking plate at the predetermined OVD). A high-speed vacuum is placed in the mouth and the mandibular movements are continued until the initial polymerization of the acrylic resin.

9) The functionally generated orthopedic occlusal device is removed from the mouth and evaluated. The path of mandibular movement seen on maxillary striking plate should also be visualized and assessed for accuracy (Figure 2.8).

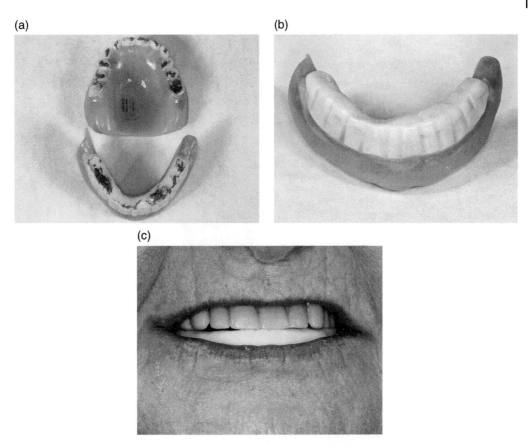

Figure 2.9 (a) Evaluation of occlusal contacts; (b) occlusal surfaces of the mandibular denture re-established; (c) occlusal device finished, polished and delivered to the patient.

If necessary, additional resin can be placed on the anterior teeth and shaped to lengthen them in case of excessive open bite. The established path of mandibular movement should not interfere with any anterior extensions. Once all resin has been added, the functionally generated orthopedic occlusal device is placed in a polymerization pressure pot containing warm water for 20 minutes and is pressurized at 20 pounds / inch [33]. Excess resin on the medial and lateral aspect of the orthopedic device is trimmed with laboratory rotary instruments. The striking plate and pin receiver are removed from the dentures, the occlusion is evaluated (using articulating film) and adjusted (using laboratory rotary instruments) as necessary to develop bilateral, stable, and harmonious contacts (Figure 2.9a). The fit and form are verified (Figure 2.9b) and the device is finished, polished and delivered to the patient (Figure 2.9c) [34]. The patient is instructed to wear the orthopedic occlusal device until stable maxillo-mandibular relationship can be recorded. The orthopedic occlusal device will help deprogram the muscles and also enable the patient to become accustomed to the new OVD.

Patients should be recalled after 2 days then every week for 4 weeks to determine if they

(a)

(b)

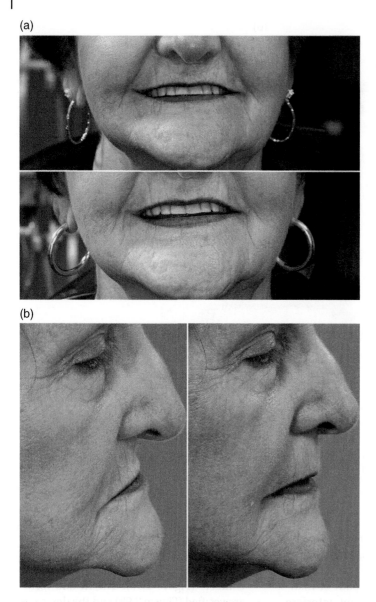

Figure 2.10 (a) Frontal view of the patient before orthopedic device (top); patient after orthopedic device (bottom); (b) profile view of the patient before orthopedic device (left); patient after orthopedic device (right). Note: there is a significant difference in OVD and esthetics before and after placement of orthopedic device.

are comfortable and can function at the established OVD. The occlusal surface can be adjusted to increase or decrease the OVD as per the requirements of each patient [13, 35, 36]. The time required for orthopedic therapy varies from patient to patient. Some patients may adapt in weeks and others may require months to adapt to the orthopedic occlusal device. Only when both the dental practitioner and the patient are satisfied with the improved esthetics, function and comfort of the patient (Figure 2.10a and b), should the next step of denture fabrication be initiated.

Summary

Identifying and registering the therapeutically and esthetically appropriate OVD and CR positions is crucial for successful denture therapy. For patients presenting with insufficient OVD, it is beneficial for both the dental practitioner and patient to evaluate the esthetics, phonetics, function, and comfort associated with planned changes in jaw relationships in the diagnostic phase before fabricating new prostheses. Acceptance of the new OVD position by both the patient and dental practitioner following an initial diagnostic phase is important for predictable and successful therapy.

This chapter describes a procedure for fabricating a diagnostic occlusal device for edentulous patients. The procedure involves modification of existing or duplicate complete dentures to evaluate proposed alterations of the existing OVD prior to fabrication of new prostheses. The primary advantage of this procedure is that functional surfaces of the occlusal device are intraorally generated by patient-induced mandibular movements limited by a central bearing device. Once completed, the orthopedic occlusal device will promote functional and muscular harmony through bilaterally balanced occlusal contacts at the desired vertical (OVD) and horizontal (CR) maxillo-mandibular relationships. The successful use of the orthopedic occlusal device aids in improvement of the health of the oral and perioral muscles and in making accurate interarch records (Neutral Zone).

References

1 Massad, J. J., Connelly, M. E., Rudd, K. D., and Cagna, D. R. (2004) Occlusal device for diagnostic evaluation of maxillomandibular relationships in edentulous patients: a clinical technique. *J Prosthet Dent*, **91**, 586–590.

2 Ismail, Y. H., George, W. A., Sassouni, V., and Scott, R. H. (1968) Cephalometric study of the changes occurring in the face height following prosthetic treatment. I. Gradual reduction of both occlusal and rest face heights. *J Prosthet Dent*, **19**, 321–330.

3 Tallgren A. (2003) The continuing reduction of the residual alveolar ridges in complete denture wearers: a mixed-longitudinal study covering 25 years. *J Prosthet Dent*, **89**, 427–435.

4 Tallgren, A., Lang, B. R., Walker, G. F., and Ash, M. M. Jr. (1980) Roentgen cephalometric analysis of ridge resorption and changes in jaw and occlusal relationships in immediate complete denture wearers. *J Oral Rehabil*, **7**, 77–94.

5 Wagner, A. G. (1989) Complete dentures with an acquired protrusive occlusion. *Gen Dent*, **37**, 56–57.

6 Niswonger, M. E. (1934) The rest position of the mandible and the centric relation. *J Am Dent Assoc*, **21**, 1572–1582.

7 Thompson, J. R. (1946) The rest position of the mandible and its significance to dental science. *J Am Dent Assoc*, **33**, 151–180.

8 Niiranen, J. V. (1954) Diagnosis for complete dentures. *J Prosthet Dent*, **4**, 727–738.

9 Porter, C. G. (1955) Cuspless centralized occlusal pattern. *J Prosthet Dent*, **5**, 313–318.

10 Boucher, C. O. (1970) *Swenson's Complete Dentures*. 6th edn. Mosby, St. Louis, MO, pp. 113–126.

11 Friedman, S. (1988) Diagnosis and treatment planning, in *Essentials of Complete Denture Prosthodontics*, 2nd edn (ed. S. Winkler). Ishiyaku Euro America, St. Louis, MO.

12 Kuebker, W. A. (1984) Denture problems: Causes, diagnostic procedures, and clinical treatment. II. Patient discomfort problems. *Quintessence Int*, **15**, 1131–1141.

13 Lyons, M. F. (1988) A review of the problem of the occlusal vertical dimension of complete dentures. *N Z Dent J*, **84**, 54–58.

14 Jeganathan, S., and Payne, J. A. (1993) Common faults in complete dentures: a review. *Quintessence Int*, **24**, 483–487.

15 Hansen, C. A. (1985) Diagnostically restoring a reduced occlusal vertical dimension without permanently altering the existing dentures. *J Prosthet Dent*, **54**, 671–673.

16 O'Grady, J. F., Reade, P. C. (1986) An occlusal splint for patients with dentures. *J Prosthet Dent*, **55**, 250–251.

17 Dabadie, M., and Renner, R. P. (1990) Mechanical evaluation of splint therapy in treatment of the edentulous patient. *J Prosthet Dent*, **63**, 52–55.

18 Palla, S. (1997) Occlusal considerations in complete dentures, in *Science and Practice of Occlusion* (ed. C. McNeill). Quintessence, Chicago, IL, pp. 457–467.

19 Rudd, K. D., and Morrow, R. M. (1985) Duplicate dentures, in *Dental Laboratory Procedures: Complete Dentures*, Vol. **1**, 2nd edn (eds K. D. Rudd, J. E. Rhoads, and R. M. Morrow). Mosby, St. Louis, MO, pp. 339–363.

20 Lindquist, T. J., Narhi, T. O., and Ettinger, R. L. (1997) Denture duplication technique with alternative materials. *J Prosthet Dent*, **77**, 97–98.

21 McHorris, W. H. (1976) TMJ dysfunction – resolution before reconstruction, in *Oral Rehabilitation and Occlusion with some Basic Principles on Gnathology*, Vol. **5** (ed. C. E. Stuart). C. E. Stuart Gnathological Instruments, Ventura, CA, pp. 151–167.

22 McHorris, W. H. (1980) Treatment of TMJ dysfunction. *J Tenn Dent Assoc*, **60**, 21–32.

23 McHorris, W. H. (1986) Centric relation: defined. *J Gnathology*, **5**, 5–21.

24 Capp, N. J. (1999) Occlusion and splint therapy. *Br Dent J*, **186**, 217–222.

25 Lang, B. R. (1994) A review of traditional therapies in complete dentures. *J Prosthet Dent*, **72**, 538–542.

26 Lytle, R. B. (1957) The management of abused oral tissues in complete denture construction. *J Prosthet Dent*, **7**, 27–42.

27 Chase, W. W. (1961) Tissue conditioning utilizing dynamic adaptive stress. *J Prosthet Dent*, **1**(11), 804–815

28 Gonzalez, J. B. (1977) Use of tissue conditioners and resilient liners. *Dent Clin North Am.*, **21**, 249–259.

29 Carlsson, G. E. (1997) Biological and clinical considerations in making jaw relation records, in *Boucher's Prosthodontic Treatment for Edentulous Patients*, 11th edn (eds G. A. Zarb, C. O. Boucher, G. E., Carlsson, and C. L. Bolender). Elsevier, St. Louis, MO, pp. 197–219.

30 Pleasure, M. A. (1951) Correct vertical dimension and freeway space. *J Am Dent Assoc*, **43**, 160–163.

31 Atwood, D. A. (1958) A cephalometric study of the clinical rest position of the mandible. Part III: Clinical factors related to variability of the clinical rest position following the removal of occlusal contacts. *J Prosthet Dent* **8**, 698–708.

32 Parr, G. R., and Loft, G. H. (1982) The occlusal spectrum and complete dentures. *Compend Contin Educ Dent*, **3**, 241–250.

33 Rudd, K. D., Morrow, R. M., and Halperin, A. R. (1985) Repairs, in *Dental Laboratory Procedures: Complete Dentures*, Vol. **1**, 3rd edn. (eds K. D. Rudd, J. E. Rhoads, and R. M. Morrow). Mosby, St. Louis, MO, pp. 383–412.

34 Zarb, G. A. (1997) Relining or rebasing of complete dentures, in *Boucher's Prosthodontic Treatment for Edentulous Patients*, 11th edn (eds G. A. Zarb, C. O. Boucher, G. E., Carlsson, and C. L. Bolender). Elsevier, St. Louis, MO, pp. 390–399.

35 Kuebker, W. A. (1984) Denture problems: causes, diagnostic procedures, and clinical treatment. II. Patient discomfort problems. *Quintessence Int*, **15**, 1131–1141.

36 Kuebker, W. A. (1984) Denture problems: causes, diagnostic procedures, and clinical treatment. III/IV. Gagging problems and speech problems. *Quintessence Int*, **15**, 1231–1238.

3

Definitive Impressions

Preimpression Considerations

A denture impression represents a negative likeness of structures within the edentulous mouth [1]. Inaccurate impressions will result in ill-fitting and unstable dentures. It is paramount that denture-bearing tissues be healthy, unchanging, and free of pathologies, soreness, inflammation, and distortion, prior to making definitive impressions. Associated systemic disease, diet, chronic trauma, and boney abnormalities should be addressed. Complete dentures fabricated on unfit denture-bearing mucosa will lead to further deterioration of tissue health and compromise the prosthetic outcome [2].

Background

Complete denture impression techniques have enjoyed a rich history in the dental literature. Three dominant philosophies have evolved: (i) definitive-pressure impressions; (ii) minimal-pressure impressions; and (iii) selective-pressure impressions [3–5]. In the first theory, attempts are made to capture the intraoral tissues in a force-loaded state. Although now seldom used for impression making, modifications of this method do have practical value as a technique for relining intaglio denture surfaces. The theory of minimal-pressure employs the principles of mucostatics. In contrast to the first theory,

attempts are designed to record the mucosa in a completely passive, nondisplaced state. Although it is impractical to achieve this fully, the use of a highly flowable impression material to avoid tissue distortion is widely accepted. The selective-pressure concept considers the variable constitution of individual intraoral anatomy and attempts to direct greater functional loads to primary stress bearing areas. Customized trays have traditionally been fashioned in which more space relief or venting is provided for the non-stress-bearing tissues. In addition, the perimeters of the trays are customized to conform to the functional extents of the vestibules. Presently, variations of this concept are the most widely accepted and practiced [6].

A majority of dental schools currently impart to students a multistep, selective, pressure technique, which includes a primary impression, construction of a primary cast on which a custom definitive impression tray is created, intraoral development of the tray periphery, and definitive impressions made with an appropriately flowable material [7, 8].

Impression Fundamentals

The following objectives for making definitive edentulous impressions have been suggested [9–11]. These axioms help to provide support, stability, and retention for the

Application of the Neutral Zone in Prosthodontics, First Edition. Joseph J. Massad, David R. Cagna, Charles J. Goodacre, Russell A. Wicks and Swati A. Ahuja.
© 2017 John Wiley & Sons, Inc. Published 2017 by John Wiley & Sons, Inc.
Companion website: www.wiley.com/go/massad/neutral

eventual prosthesis and preserve health of the biologic tissues:

1) The impression should cover the entire basal seat within the limits of function of various orofacial muscles. Maximum coverage helps dissipate the forces over a large area, reducing the amount of force on each square millimeter.
2) The impression should have maximum contact with denture-bearing mucosa, to ensure adequate fit of the final prosthesis.
3) The borders of complete dentures should be defined physiologically so that they are in harmony with the anatomic and functional limits of the denture foundation and adjacent tissues.
4) The impression should be made such that its dimensions and contours replicate the intended contours of the definitive complete denture.

In order to make an accurate impression, it is imperative to understand the anatomy and physiology of the individual edentulous mouth and the methods to identify and capture dynamic postures. The vestibular extents of both arches should be relieved and functionally re-established by associated muscular activation within the impression to form the eventual denture flange borders. A medium that has both additive and subtractive working properties is preferred to accomplish this. Care must be taken not to suppress or overextend areas of attachment such as the frenulae or areas that express physiologic mobility requirements such as the floor of the mouth and retromylohyoid region of the mandible [12]. The postpalatal seal area is defined as the soft tissue area at or posterior to the junction of the hard and soft palates, on which pressure within physiologic limits can be applied by a maxillary complete denture to aid in its retention. It is important to identify, assess, and capture the extent of this region in the maxillary impression [13].

It is also crucial to understand where more moderate pressure can be applied and where it should be minimized during the making of definitive impressions [14]. In the edentulous

maxilla, moderate pressure may be applied on the horizontal portions of the hard palate, tuberosities, rugae, and the ridge crests (if covered with firm, healthy, nondisplaceable mucosa). These are considered to be stress-bearing areas in the maxilla. The median palatal raphe and incisal papilla need to be relieved and recorded with minimum pressure. For making an impression of the edentulous mandible, the buccal shelves, ridge crests, and retromolar pads (if firm) are considered stress-bearing areas and can be recorded with moderate pressure. Mobile mandibular ridge crests and associated lingual extensions need to be recorded with less pressure.

Impression Materials

Historically, much attention has been given to the wide variety of materials available for making edentulous impressions including plaster, modeling plastic impression compound, zinc oxide-eugenol paste, irreversible hydrocolloid, polysulfide, polyether, and vinyl polysiloxane. All of these materials may perform adequately in the hands of an experienced practitioner.

More recently, vinyl polysiloxane (VPS) has become a popular material of choice for definitive dental impressions. VPS impression material, an addition reaction silicone, is available in different viscosities (extra light, light, medium, heavy, and rigid) and with different working times. Subsequent layers of VPS can be added sequentially to existing polymerized material, permitting a layering technique. Heavier bodied materials offer sufficient working times and viscosity to permit physiological molding of the borders of the impression tray and primary stress-bearing areas, while more flowable materials can capture mobile tissues without distortion [15]. Vinyl polysiloxane impression materials are easy to use and disinfect, are accurate, have favorable tear strength, and can remain dimensionally stable for an extended period. These materials have excellent elasticity and wettability, which enhance detailed recording of oral tissues and complement gypsum cast production [16]. They also offer the benefit of repeated pourability for the production of multiple casts from a single impression.

Edentulous Impression Trays

Physical properties of the impression tray and its manipulation constitute important considerations in the impression process. Historically, the use of custom impression trays to make definitive edentulous impressions has been considered essential for accurate results. Improvements in impression materials and impression trays have made it possible to avoid the necessity for custom impression trays [15]. Today, stock trays are available that can be molded and shaped to conform to the dimensions and contours suitable to the anatomy of edentulous patients. These can eliminate the need for making primary impressions and for construction of custom trays. Stock impression trays used for making a definitive impression, should meet the following requirements [15]:

1) The tray should be rigid enough to permit border-molding procedures and support the impression material.
2) The trays should be available in various sizes for variety of edentulous arches. A large tray used on a small arch or small tray used on a large arch will distort the tissues.
3) The trays should permit both subtractive and additive modification of denture borders to permit recording the vestibular extensions physiologically without distorting or displacing soft tissues.
4) The tray design should provide retention for the impression material.
5) The tray handle should not interfere with border-molding movements.

Specialized thermoplastic stock trays, specifically designed to meet the requirements of edentulous patients, have been produced for making definitive impressions (Massad edentulous impression trays). These trays are constructed of a polystyrene-based polymer and can be customized and molded according to individual patient anatomy and physiology. They are designed with contoured vestibular borders, retention slots, and ergonomic finger rests. The retention slots designed in the tray help record the denture-bearing soft tissues with minimum distortion and displacement and also provide a means of mechanical retention of impression material. Ergonomic finger-rest extensions help the molded tray to remain rigid and to distribute pressure evenly across the denture bearing areas. Well-designed tray handles are positioned such that they do not interfere with the border-molding movements.

Technique for Making Single Appointment Definitive Impressions for Conventional Complete Dentures

An innovative technique is presented to simplify the procedure of impression making and decrease the number of clinical appointments [17].

Tray Selection and Tray Adaptation

As a preliminary step, before engaging the mouth, it is recommended to clean and hydrate the denture-bearing soft tissues prior to impression making. This can be accomplished by asking the patient to take a few sips of warm water, to hold it in the mouth and then to use a rotary oscillating or pulsating toothbrush to massage the denture bearing soft tissues. Care should be taken not to allow excess water to leak out of the mouth prior to rinsing. This procedure will not only help massage and hydrate the tissues but will

Figure 3.1 Maxillary thermoplastic tray immersed in water bath.

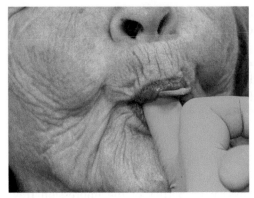

Figure 3.2 Patient performing various movements to physiologically mold thermoplastic tray.

also help decouple residual plaque and any residual denture adhesive that may compromise the tray or impression procedures.

The featured trays are available in five maxillary sizes and five mandibular sizes. The appropriately sized stock impression tray is selected by placing each end of a measuring caliper on the posterior third (maxillary tuberosity for the maxilla and retromolar pad for the mandible) of the alveolar ridge. This measurement should correlate with measured ridge size. The selected tray is tried into the mouth and verified. The entire tray can be molded by heat to improve its conformation to the patient's arch by placing into a controlled heated water bath at 165 °F for approximately 20–30 s or until the tray material softens (Figure 3.1). Cheek retractors can be used to permit easy access to the oral cavity. After softening, the tray is immediately reinserted into the mouth and the patient guided to perform border-molding movements (Figure 3.2).

The tray material hardens in 5–10 s, after which it can be removed from the mouth and analyzed for accuracy. The tray and tray borders can be modified using laboratory rotary instruments and direct flame heat molding if necessary. Ideally, the tray borders should be positioned in the middle of the vestibules and be wide enough not to create tissue impingement.

Fabrication of Tray Stops

Making a definitive impression requires multiple placements of the impression tray in the oral cavity. Formation of tray stops is critical to this technique so that consistent and repeatable tray placements position can be achieved. Tray stops help accomplish several objectives as described in Box 3.1 [15].

Nickel-sized circles of rigid-viscosity VPS impression material are injected into the maxillary tray in four locations (incisor, molar and mid-palate) and the tray is carried into the patient's mouth. For the mandibular tray, stops are formed in three locations (incisor and bimolar). Trays are centered and seated on the respective

---

Box 3.1 Objectives of tray stops

- To provide a constant path of tray insertion.
- To provide appropriate space for impression material.
- To help stabilize the tray in a centered position and prevent rotational movement of the tray during impression making.
- To help stabilize the tray during patient generated border molding procedures.
- To prevent overseating of the tray while making a definitive impression.

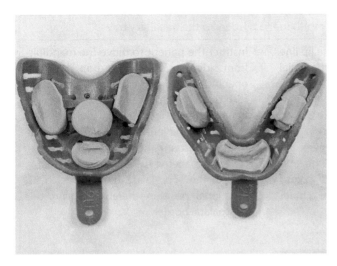

Figure 3.3 Tray stops formed in maxillary and mandibular trays.

arch and kept 2–3 mm away from the vestibular sulcular extents. Impression material is allowed to polymerize for 2 minutes. The tray is removed, evaluated and modified as needed before proceeding. The tray stops are trimmed with a sharp blade to reduce areas of excessive tissue contact (Figure 3.3).

Border Molding the Impression Tray

Border molding is the process of "shaping of the periphery of an impression tray, by functional or manual manipulation of the adjacent tissue, to duplicate the contour and size of the vestibule." The border-molding procedure aids in defining the appropriate length and dimension of denture border, which is critical to establish a peripheral seal. Overextended denture borders may encroach on muscle attachments leading to soreness, ulceration of the tissues, and denture dislodgement during function. Rigid viscosity VPS is dispensed along the peripheral extents of the tray (Figure 3.4), placed in the mouth and centered over the ridge with the aid of the tray stops.

The dynamic manipulations for physiologically molding the borders of the maxillary tray are listed in Box 3.2. The tissue

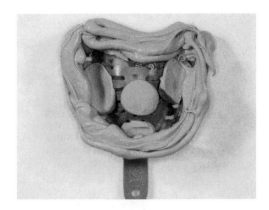

Figure 3.4 VPS material dispensed along peripheral extent of maxillary tray.

manipulations for physiologically molding the borders of the mandibular tray are listed in Box 3.3.

Following complete polymerization of the impression material, the impression tray is removed from the mouth and the borders evaluated to ensure recording of all the anatomic and functional detail. Areas of tray show through should be adjusted with laboratory rotary instruments or a sharp blade by 1–2 mm. Additional relief should be provided for non-pressure bearing tissues. All the borders are then relieved by 1–2 mm

> **Box 3.2 Dynamic manipulations for molding the borders of the maxillary tray**
>
> 1) The philtrum is grasped close to the lip line and pulled downward to record the labial frenum.
> 2) The patient is asked to purse the lips outward with a sucking action (Figure 3.5) to register the labial vestibular extension.
> 3) The corners of the mouth and cheek are grasped with forefingers and thumb and pulled downward and forward to record the buccal frena and buccal vestibular extensions.
>
> 4) Instruct the patient to move the mandible from side to side and to drop down to opening position. Both these actions help record the disto-buccal vestibular sulci and hamular frena (Figure 3.6).
> 5) Occlude the patient's nostrils and ask them to exhale through the nose only (Figure 3.7) (Valsalva maneuver). This will lead to migration of the soft palate downwards to its functional position thereby forming the postpalatal border.

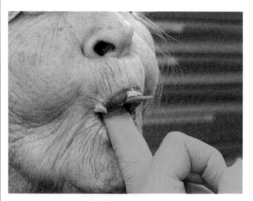

Figure 3.5 Patient purses lips outward with a sucking action to register the labial vestibular extension.

Figure 3.6 Patient moves the mandible from side to side, drops it down to opening position, to record distobuccal vestibular sulci and hamular frena.

Figure 3.7 Patient's nostrils occluded and patient instructed to exhale only through nose to aid in forming postpalatal border.

Box 3.3 The tissue manipulations for physiologically molding the borders of the mandibular tray

1) The patient is asked to place the tip of the tongue straight out and forward, then side to side, then back touching the roof of the mouth, and then to swallow. These movements help record the lingual vestibule and retromylohyoid fossae without overextension of borders.

2) The lower lip is grasped at the lip line, then pulled upward and forward; this registers the labial frenum and labial vestibular extensions.

3) The tray is secured by placing two fingers on the tray finger support and the thumb under the patient's chin (Figure 3.8). The patient is asked to purse the lips out and suck. This records the buccal and labial vestibules.

4) The corners of mouth are grasped with the forefingers and thumb and pulled downward and forward to record the buccal frena and distobuccal vestibular extensions.

5) The patient is asked to open wide and then close the lower jaw on the practitioner's finger while the practitioner exerts a downward force on the tray (Figure 3.9). The activation of the masseter muscle causes bulging of the buccinator, leading to the development of the masseric notch in the impression. This also helps define the posterior extent and retromolar pad areas.

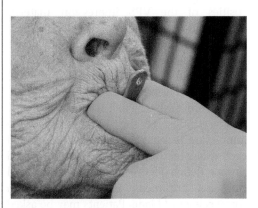

Figure 3.8 Impression tray is secured and patient is asked to purse lips out and suck to record the buccal and labial vestibules.

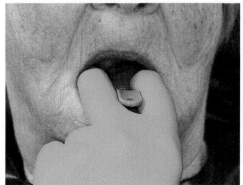

Figure 3.9 The patient asked to open wide and then close the lower jaw while the practitioner exerts a downward force on tray to define posterior extent and retromolar pad areas.

prior to making definitive impressions (Figure 3.10).

Final (Definitive) Impression

The appropriate viscosity of VPS for the final impression is chosen based on the condition of the existing denture-bearing tissues (tissue character and mobility.) Medium / light viscosity VPS is recommended for firm, nondisplaceable tissues. For mobile, flabby or redundant tissues, relief can be provided by removing adjacent set impression materials from the tray and extra-light viscosity VPS can be used. A combination of different viscosities can be used simultaneously, depending on the nature and character of denture-bearing tissues.

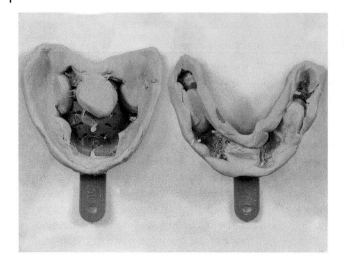

Figure 3.10 Border molded and relieved maxillary and mandibular impression tray.

Figure 3.11 Maxillary impression tray loaded with two different viscosities of VPS material.

The tray is loaded with the chosen viscosities of the VPS impression material (Figure 3.11), and is placed and centered in the mouth until tactile resistance is felt once the established tray stops are approximated. The manipulations listed previously for the respective arches are repeated to again mold and record all borders. Following complete polymerization of the VPS material, the impression is removed from the mouth and each evaluated for anatomic, functional and surface detail (Figure 3.12). Excess material can be trimmed, if necessary, with sharp scissors / blade. Patients may be asked to remove the impression from the mouth

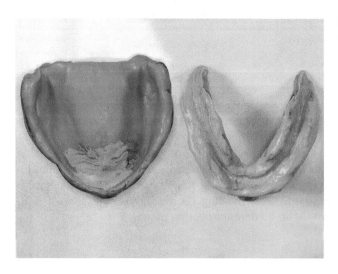

Figure 3.12 Maxillary and mandibular definitive edentulous impressions.

themselves so that they can truly perceive the retention of their new prosthesis. If patients seem to be dissatisfied with the retention of the impression, they should be advised to consider an implant-retained prosthesis.

Techniques for Making Single Appointment Definitive Impressions for Implant-Assisted Complete Dentures and Immediate Dentures

The use of endosseous dental implants to assist in the support, stability, and retention of removable prostheses is considered an effective treatment modality for edentulous patients. Individuals wearing implant-assisted overdentures typically report improved oral comfort and function when compared with conventional, mucosa-supported prosthesis [18–23]. Vinyl polysiloxane (VPS) impression materials are well suited to address both the accurate registration of denture-bearing tissues and peripheral anatomy, and the stable recording of dental implant positions and individual implant trajectories [24].

Implant overdenture VPS impression techniques involve overdenture attachment selection, tray selection and adaptation, tray stops, border molding, and the definitive impression.

Attachment Selection

A distinction should be made regarding the number of implants and configuration of the attachment systems recommended for the individual edentulous patient. The number of implants could range from two to four (accompanied by free-standing abutments, which retain the denture but still rely on the mucosa contacting the denture base to help support the prosthesis), to four or more (using a bar constructed to connect the abutments to provide both retention and support for the prosthesis).

Implant-Retained Overdentures

A variety of attachment systems have been developed and marketed for use with implant-retained overdentures. Some systems are designed to pick up retentive components directly in the finished denture. Others use an indirect approach, which include an impression with transfer copings (Figure 3.13).

A master cast is created that displays all of the anatomy used to construct an overdenture and also contains analogs to which retentive components are attached and processed into the denture base. Impressions for implant retained, free-standing, attachment systems are generally made using a "closed-tray" technique.

Tray Selection

Carefully examine the dimensions of the dental arch and select the appropriate stock impression tray. The thermoplastic impression trays illustrated (strong-Massad dentate and implant trays – Figure 3.14) are constructed from a polystyrene-based polymer and are available in three maxillary sizes and three mandibular sizes: small, medium, and large. These trays are clear to permit see-through visibility to assist in size selection, and heat moldable to aid in adapting the tray. Sufficient room between the tray and all implant attachments and impression components must be provided.

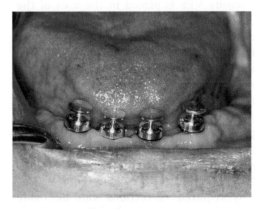

Figure 3.13 Implant attachment transfer copings connected to each overdenture abutment.

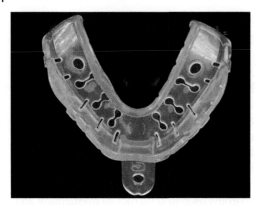

Figure 3.14 Clear thermoplastic stock impression tray for implant / dentate impressions.

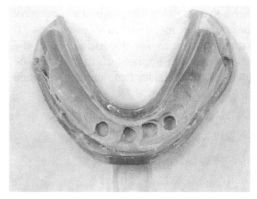

Figure 3.15 Tray stops formed with rigid VPS impression material.

Impression Technique

To achieve consistently repeatable tray placements, tray stops are developed as described before using a rigid-viscosity material (Figure 3.15). Border molding using a rigid-viscosity material is accomplished using the listed protocol (Figure 3.16). Once the material has set, the tray is removed from the mouth and the borders are adjusted and relieved 1–2 mm, in preparation of the definitive impression. Care should be taken to also relieve any set material that has engaged the implant attachment transfer components.

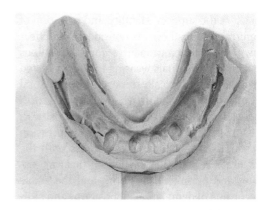

Figure 3.16 Tray borders molded and trimmed prior to making final impression.

Both low and high viscosity VPS materials are placed in the developed tray. High viscosity VPS is placed in the area of the abutment impression copings. Lower viscosity materials are used elsewhere in the tray. Low-viscosity VPS material is injected around the implant attachment transfers in the mouth. The tray is placed into the mouth using tray stops and tray borders to guide tray placement. The border molding protocol is repeated. Once the material has polymerized, the tray is removed and the impression examined for correct anatomic, functional and surface detail (Figure 3.17). Any required corrections can be made using the subtractive and additive methods described above.

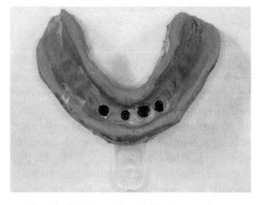

Figure 3.17 Mandibular closed tray definitive impression for implant retained overdenture fabrication.

Implant-Retained and Supported Overdentures

Support and retention for a complete implant overdenture is possible with enough adequately dispersed free-standing abutments. However, this section will discuss impression procedures unique to bar-supported prostheses [25].

A variety of manufacturers offer transfer copings, which are attached by screws to the intraoral components, for use during impression procedures. Transfers can be made directly from the implant interfaces or from the interfaces of intervening transmucosal abutments. Resultant master casts will contain analogs of the construction platforms for a connecting bar, which will be complemented by retentive components processed within the overdenture. Impressions for implant-bar retained and supported overdentures are generally made using an "open-tray" technique. Previously mentioned edentulous or dentate thermoplastic trays can be used. Direct transfer copings are connected intraorally to each implant or abutment interface (Figure 3.18).

Positions are estimated and holes placed in the tray with an acrylic bur to allow passage of the transfers through the tray when seated. Tray stops are accomplished as before, using rigid viscosity VPS material. It is important to relieve the areas of any set material adjacent to the transfer copings (Figure 3.19).

Border molding is accomplished as before with rigid-viscosity material. Low-viscosity material is injected around each attached correct spacing viscosity material placed into the tray except around the implants where a rigid-viscosity material is injected in the tray. Once properly seated in the mouth, excess material extruding through the holes created in the tray should be cleared quickly to visualize and access the attachment screws easily. Upon polymerization of the VPS material, loosen each of the attachment screws and then remove the impression, including the transfer copings, from the mouth. The definitive impression should be inspected for proper containment of the copings and appropriate anatomic, functional, and surface details (Figure 3.20). Analogs for the construction

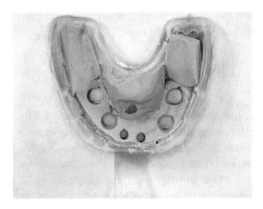

Figure 3.19 Tray stops formed and tray is relieved in areas adjacent to transfer copings.

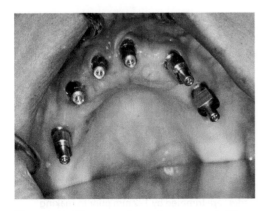

Figure 3.18 Direct transfer copings connected to all implants.

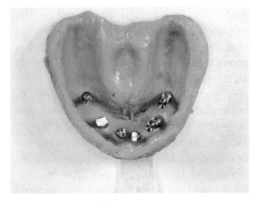

Figure 3.20 Open tray definitive impression for fabrication of implant supported and retained overdenture.

(a)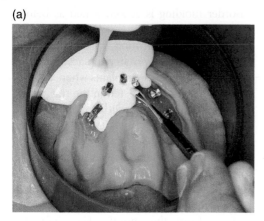

(b)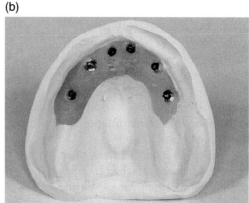

Figure 3.21 (a) Implant analogs attached to transfer copings and soft tissue material placed appropriately prior to cast fabrication; (b) definitive implant level soft tissue cast.

platforms for the bar are attached to the transfers within the impression and surrounding soft tissue material placed prior to cast production (Figures 3.21a and 3.21b).

Immediate Dentures

Combined with the many advantages of VPS impression material, a convenient definitive immediate denture impression technique can be accomplished in a single appointment [26]. The appropriate size dentate, clear, thermoplastic tray is selected and adapted to the partially dentate arch. Tray stops are created for the existing residual teeth and ridge using rigid-bodied VPS (Figure 3.22). Rigid-viscosity materials are placed on the perimeter and the adapted tray is seated using the stops as guides. Border molding is accomplished using the methods previously described (Figure 3.23).

For partially dentate patients, using an elastic material for this process allows the impression to be removed from the mouth despite engaging potentially confounding tissue undercuts. Upon removal, the borders are defined and relieved. The definitive impression is made using low viscosity VPS material in the edentulous areas and injecting extra-low viscosity material around all residual teeth

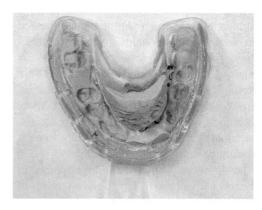

Figure 3.22 Tray stops formed for immediate denture definitive impression.

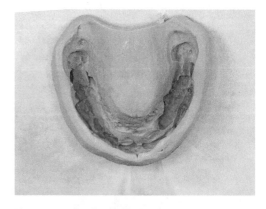

Figure 3.23 Border molded impression. Note: borders are trimmed by 1–2 mm prior to making definitive impressions.

using manual syringes (Figure 3.24). Extra-low viscosity VPS material possesses lower tear strength, permitting easier recovery of the cured impression from the patient's mouth. Its use also minimizes the chances of breakage of teeth (especially those that are prone to breakage, which exhibit long clinical crowns due to advanced periodontal disease) from the cast, while recovering the cast from the impression.

The loaded tray is placed into the mouth using the stops as guides. All border-molding manipulations are repeated. Upon polymerization of the VPS, remove and inspect the impression for appropriate anatomic, functional, and surface details (Figure 3.25).

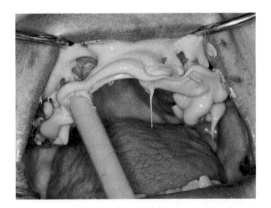

Figure 3.24 Extra low viscosity impression material injected around all teeth.

Master Cast Production

Once satisfied with the definitive impressions, recommended disinfection procedures are performed. Although VPS materials possess excellent dimensional stability and can generally be transported dependably, some practitioners prefer to verify quality by creating master casts in office. Definitive casts should be accurate, void free and created with a properly formed base [27]. A base former can be used with a bulk silicone product to customize (bead and box) the outline and thickness of the cast base, adapted to the appropriate dimensions. An approved, ADA standard 25, type-III stone may be used for construction (Figure 3.26). The process of mixing these powdered gypsum products with the appropriate water is critical to the cast outcome. Automatic dispensing and mixing machines (Figure 3.27) are useful to control consistent production. Vacuum mixing at reduced atmospheric pressure helps to prevent void formation. The resulting cast should require little trimming or surface preparation (Figure 3.28).

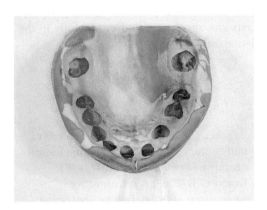

Figure 3.25 Definitive impression for fabrication of maxillary immediate denture.

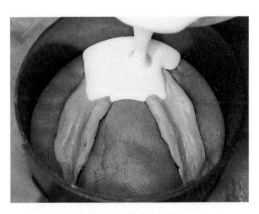

Figure 3.26 Fluid dental stone poured in boxed and beaded impression for definitive cast fabrication.

Figure 3.27 Automatic dispensing unit and automatic vacuum mixing unit for dental stone.

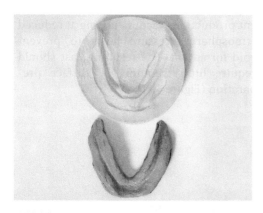

Figure 3.28 Definitive mandibular cast (top) and definitive mandibular impression (bottom).

Summary

This chapter describes a novel technique of making a definitive impression in a single appointment eliminating the need of custom tray fabrication. Contemporary materials and trays have facilitated the development of this technique. This technique is simple, easy, accurate, and can be incorporated into any dental practice. Quality master casts can also be created successfully in the dental office.

References

1 The glossary of prosthodontic terms (2005). *J Prosthet Dent*, **94**, 44.
2 Kuebker, W. A. (1984) Denture Problems: Causes, diagnostic procedures, and clinical treatment. I. Retention problems. *Quintessence Int*, **10**, 1031–1044.
3 Addison, P. I. (1944) Mucostatic impressions. *J Am Dent Assoc*, **31**, 941–946.
4 Boucher, C. O. (1951) A critical analysis of mid-century impression techniques for full dentures. *J Prosthet Dent*, **1**, 472–491.
5 Drago, C. J. (2003) A retrospective comparison of two definitive impression techniques and their associated post insertion adjustments in complete denture prosthodontics. *J Prosthod*, **12**, 192–197.

6 Felton, D. A., Cooper, L. F., and Scurria, M. S. (1996) Predictable impression procedures for complete dentures. *Dent Clin North Am*, **40**, 39–51

7 Petropoulos, V. C., and Rashedi, B. (2003) Current concepts and techniques in complete denture final impression procedures. *J Prosthodont*, **12**, 280–287.

8 Arbree, N. S., Fleck, S., and Askinas, S. (1996) The results of a brief survey of complete denture prosthodontic techniques in predoctoral programs in North American dental schools. *J Prosthodont*, **5**, 219–225.

9 DeVan, M. M. (2005) Basic principles in impression making. *J Prosthet Dent*, **3**, 503–508. Originally published in 1952.

10 Ivanhoe, J. R. (2009) Final impressions and creating the master casts, in *Textbook of Complete Dentures*, 6th edn. (eds. A. O. Rahn, J. R. Ivanhoe, K. D. Plummer). People's Medical Publishing House, Shelton, CT, pp. 105–128.

11 Jacob, R. F., Zarb, G. (2012) Maxillary and mandibular substitutes for the denture-bearing area, in *Prosthodontic Treatment for Edentulous Patients: Complete Dentures and Implant-Supported Prostheses*, 13th edn. (eds. G. Zarb, J. A. Hobkirk, S. E. Eckert, R. F. Jacob). Elsevier Inc, St Louis, MO, pp. 161–179.

12 Chaffee, N. R., Cooper, L. F., and Felton, D. A. (1999) A technique for border molding edentulous impressions using vinyl polysiloxane material. *J Prosthod*, **8**, 129–134.

13 Ansari, I. H. (1997) Establishing the posterior palatal seal during the final impression stage. *J Prosthet Dent*, **78**, 324–326.

14 Chopra, S., Gupta, N. K., Tandan, A., *et al.* (2016) Comparative evaluation of pressure generated on a simulated maxillary oral analog by impression materials in custom trays of different spacer designs: An in vitro study. *Contemp Clin Dent*, **7**, 55–60.

15 Massad, J. J., and Cagna, D. R. (2007) Vinyl polysiloxane impression material in removable prosthodontics. Part 1: edentulous impressions. *Compend Contin Educ Dent*, **28**, 452–459.

16 Norling, B. K., and Reisbick, M. H. (1979) The effect of nonionic surfactants on bubble entrapment in elastomeric impression material. *J Prosthet Dent*, **42**, 342–347.

17 Massad, J. J., Lobel, W., Garcia, L. T., *et al.* (2006) Building the edentulous impression – a layering technique using multiple viscosities of impression materials. *Compend Contin Educ Dent*, **27**, 446–451.

18 Geertman, M. E., Boerrigter, E. M., Van't Hof, M. A., *et al.* (1996) Two-center clinical trial of implant-retained mandibular overdentures versus complete dentures-chewing ability. *Community Dent Oral Epidemiol*, **24**, 79–84.

19 Geertman, M. E., van Waas, M. A., van't Hof, M. A, and Kalk, W. (1996) Denture satisfaction in a comparative study of implant-retained mandibular overdentures: a randomized clinical trial. *Int J Oral Maxillofac Implants*, **11**, 194–200.

20 Kapur, K. K., Garrett, N. R., Hamada, M. O., *et al.* (1999) Randomized clinical trial comparing the efficacy of mandibular implant-supported overdentures and conventional dentures in diabetic patients. Part III: comparisons of patient satisfaction. *J Prosthet Dent*, **82**, 416–427.

21 Raghoebar, G. M., Meijer, H. J., Stegenga, B., *et al.* (2000) Effectiveness of three treatment modalities for the edentulous mandible. A five-year randomized clinical trial. *Clin Oral Implants Res*, **11**, 195–201.

22 Awad, M. A., Lund, J. P., Dufresne, E., and Feine, J. S. (2003) Comparing the efficacy of mandibular implant-retained overdentures and conventional dentures among middle-aged edentulous patients: satisfaction and functional assessment. *Int J Prosthodont*, **16**, 117–122.

23 Awad, M. A., Lund, J. P., Shapiro, S. H., *et al.* (2003) Oral health status and treatment satisfaction with mandibular implant overdentures and conventional dentures: a randomized clinical trial in a senior population. *Int J Prosthodont*, **16**, 390–396.

24 Massad, J. J., and Cagna, D. R. (2007) Vinyl polysiloxane impression material in removable prosthodontics. Part 3: implant and external impressions. *Compend Contin Educ Dent*, **28**, 554–560.

25 Massad, J. J., Cagna, D. R., Wicks, R. A., and Selvedge, L. A. (2016) Cameograms: A new technique for prosthodontic applications. *Dent Today*, **35**, 80, 82, 84–85.

26 Massad, J. J., and Cagna, D. R. (2007) Vinyl polysiloxane impression material in removable prosthodontics. Part 2: immediate denture and reline impressions. *Compend Contin Educ Dent*, **28**, 519–526.

27 Rudd, K. D., Morrow, R. M., and Bange, A. A. (1969) Accurate casts. *J Prosthet Dent*, **21**, 545–554.

4

Fabricating Record Bases, Occlusal Rims, and Mounting a Central Bearing Device

Introduction

Following the fabrication of the definitive casts, laboratory procedures are indicated in preparation for the next clinical appointment (centric relation records). Record bases and occlusal rims are fabricated for the maxillary and the mandibular casts to record the esthetic information of the maxillary arch and neutral zone of the mandibular arch respectively, and to register centric relation.

Fabrication of Record Base and Occlusal Rims

A critical aspect of diagnosis and rehabilitation of the edentulous mouth is accurate registration of the centric relation jaw relationship [1]. The accuracy of this registration is directly related to the retention, support, and stability of the record base [2–4]. Loose and ill-fitting record bases are unstable and have a tendency to displace and unseat in the oral cavity, resulting in potential occlusal interferences and malocclusion [2, 4, 5]. Accuracy of a record base is dependent on the accuracy of the impression and the definitive cast, and the material and technique utilized for its fabrication [6].

Numerous materials and laboratory techniques for the construction of accurate and stable record bases have been suggested,

including: a sprinkle-on technique using acrylic resin [6, 7], manual adaptation of acrylic resin [6], matrix adaptation of acrylic resin, thermoplastic adaptation of shellac/hard baseplate wax [6], vacuum adaptation of thermoplastic materials [8], fluid resin processing, heat-polymerized compression/injection molding of acrylic resin [9], use of resilient liner [2]/nonresilient liner with shellac/acrylic resin [5], and use of final impressions as a preliminary record base [10, 11]. Currently, the most widely used material for the fabrication of record bases is autopolymerized/light polymerized acrylic resin. Record bases may be handmade, milled, or digitally printed (Figure 4.1), as discussed in detail in Chapter 11. This chapter will detail the fabrication of handmade record bases using light polymerized acrylic resin.

Two sets of maxillary and mandibular record bases are fabricated for the patient. The first set is used for recording the esthetic information of the maxillary arch and the neutral zone of the mandibular arch. The second set of record bases is used for registering the centric relation. The second set is designed to be slightly underextended, to minimize unwanted movement of the record bases during muscular/vestibular movements. The fabrication of the light polymerized record base is described below:

1) The casts are evaluated for areas that are undercut relative to the planned removal path of the record base. The undercut areas are blocked out using standard base

Application of the Neutral Zone in Prosthodontics, First Edition. Joseph J. Massad, David R. Cagna, Charles J. Goodacre, Russell A. Wicks and Swati A. Ahuja.
© 2017 John Wiley & Sons, Inc. Published 2017 by John Wiley & Sons, Inc.
Companion website: www.wiley.com/go/massad/neutral

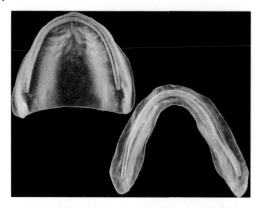

Figure 4.1 Digitally printed record base. Maxillary record base (left). Mandibular record base (right). Note: retention ledge incorporated in the design for aiding with the retention of the wax/prosthetic teeth to the record base.

plate wax. Minimal block-out is desirable, as excess block-out may result in record bases that demonstrate poor fit and compromise retention and stability during clinical procedures. Areas that typically require block-out include:

- labial aspect of maxillary anterior ridge;
- buccal aspect of tuberosities;
- labial aspect of mandibular anterior ridge;
- lingual aspect of mandibular posterior ridge;
- around prominent rugae;
- around prominent frenula; and
- around prominent and sharp mandibular ridge crests.

2) A separating medium (silicone-based jelly or alginate-based tin-foil substitute) is applied to the cast following the manufacturer's recommendations. This will permit easy separation of the record base from the cast with minimal risk of damaging the cast.

3) A single sheet of light activated resin is adapted to the cast. It is recommended to cut the light activated resin sheet in to two halves to facilitate cross arch adaptation of material for the mandibular cast. Excessive pressure should not be applied to the resin sheets as this unduly thins the resin and weakens the record bases.

4) The excess material is trimmed using dull instruments to avoid scoring the cast. The

adaptation of the resin to the cast is evaluated and corrected where required.

5) The cast is placed in a light polymerizing unit. The timer is adjusted and set according to manufacturer's recommendations to permit complete polymerization of the material.

6) The cast is removed from the light polymerizing unit and the record base is carefully retrieved from the cast and placed by itself in the light polymerizing unit (with the intaglio surface facing upwards) to permit complete polymerization of the intaglio surface of the record base. The record base is removed from the light polymerizing unit and inspected to assure accurate reproduction of tissue detail and contours.

7) Using rotary instrumentation and burs, excess material is trimmed from periphery of record bases. The borders and tissue surfaces should be inspected for sharp areas and / or ridges, which may traumatize the patient on placement, and corrected as needed. Finally, the record base is placed on the cast and carefully examined for proper fit.

Fabrication of Maxillary Wax Occlusal Rim

A wax occlusal rim is added to the maxillary record base to aid in evaluation and registration of optimal denture tooth position, occlusal plane, and jaw relations. Measurements made during initial patient examination (diagnosis as discussed in Chapter 1) can be used to customize the position and contour of the wax occlusal rim. The technique is described below:

1) The lip ruler is used to measure the distance between the premaxillary ridge crest and upper lip at rest and during smiling (esthetic space) during the diagnostic appointment (Figure 4.2a).

2) This measurement is transferred on the cast using the lip ruler and a compass

(a)

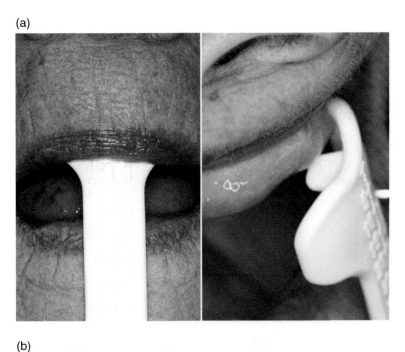

(b)

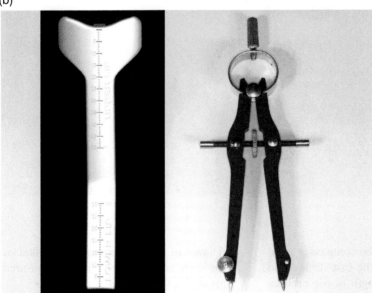

Figure 4.2 (a) Esthetic space measured and recorded during diagnostic appointment (left) The tab on the lip ruler placed on the crest of the ridge while measuring esthetic space (right); (b) lip ruler (left) and compass (right); (c) the lip ruler is appropriately positioned on the maxillary cast and the esthetic space measurement transferred using the compass; (d) the planned incisal length of maxillary occlusal rim; (e) the compass is rotated laterally on the cast to scribe a line recording the esthetic space measurement on cast.

(c)

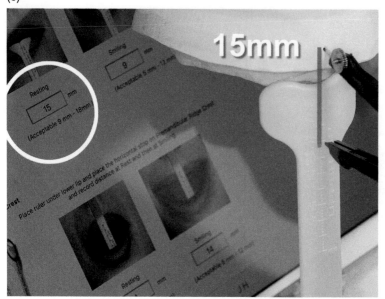

(d)

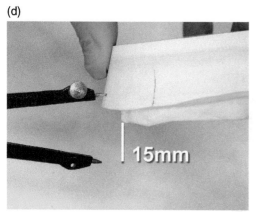

(e)

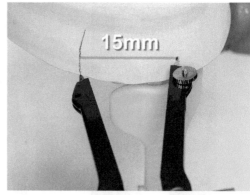

Figure 4.2 (Continued)

(Figures 4.2b and c). The compass is set at an arbitrary point on the cast surface and the planned incisal length of the occlusal rim is evaluated (Figure 4.2d). Next, it is rotated to the lateral surface of the cast, a line is then scribed on the base of the cast to record the set measurement (Figure 4.2e).

3) The ridge area may be roughened with laboratory rotary instruments or a retention ledge may be incorporated in the design of the record base (Figure 4.3) for improving the retention of the wax and prosthetic teeth to the record base.

A layer of sticky wax may be applied to the ledge area or the roughened ridge area for aiding the retention.

4) A single sheet of baseplate wax is warmed and folded into thirds. The wax is adapted on the record base along the edentulous ridge.

5) The rim is further secured to the record base by using a heated #7 wax spatula.

6) The rim is contoured according to the arch form. The average distance between the most anterior point of the wax occlusal rim and the most posterior point of the

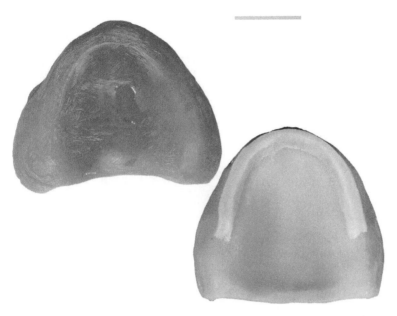

Figure 4.3 The ridge areas roughened with laboratory rotary instruments (left). Retention ledge incorporated in the design (right).

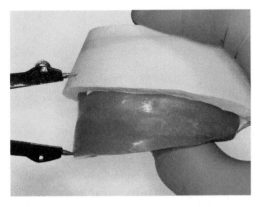

Figure 4.4 The compass is repositioned on cast and the predetermined length marked on rim.

incisive papilla should be approximately 12 mm [12]. The anterior aspect of the wax rim should have a labial inclination (10–15°) [13], which should be more pronounced for female patients as compared to male patients.

7) The compass is repositioned on the cast to evaluate the length of the anterior wax occlusal rim. A mark representing the predetermined length is scribed on the wax occlusal rim (Figure 4.4). The rim

is decreased or increased in length accordingly.

8) A heated rim former can be used to decrease the length of the rim. The elevated ridge of the rim former is positioned in the hamular notches of the cast (Figure 4.5a). The rim former is rotated towards the cast, melting the rim to the mark representing the planed anterior rim length (Figure 4.5b). The posterior height of the rim and the plane of occlusion are produced arbitrarily using this technique.

9) The occlusal rim is contoured to be 8–10 mm in width in the posterior region and 6–8 mm in width in the anterior region.

10) The occlusal rim is smoothened and further contoured in preparation for clinical use (Figure 4.6).

A maxillary wax rim, which is fabricated and adjusted clinically according to patient's esthetics and phonetics, serves as a blueprint for clinically determined esthetic information for transfer to the laboratory. The name "esthetic blueprint" is therefore appropriate

(a) (b)

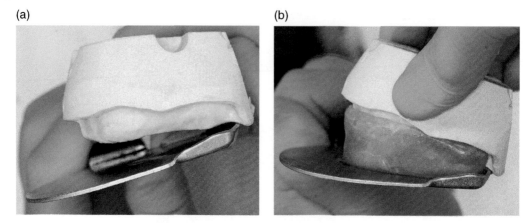

Figure 4.5 (a) Elevated ridge of the rim former is positioned in hamular notches of the cast and it is rotated towards cast while decreasing height of wax occlusal rim; (b) the heated rim former used to produce the desired height of wax occlusal rim.

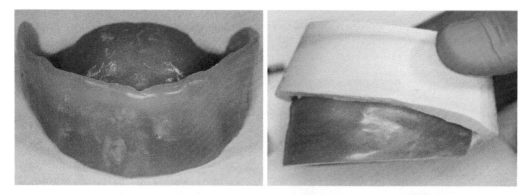

Figure 4.6 The completed maxillary wax occlusal rim frontal (left) and lateral view (right).

for the maxillary wax occlusal rim. Facial veneer shell anterior teeth can be waxed to the rim for esthetic assessment at the next clinical appointment (described in detail in Chapter 5).

Fabrication of Neutral Zone Mandibular Occlusal Rim

Optimal denture tooth position in the buccolingual dimension for the mandibular arch can be best accomplished by registration of the physiologic neutral zone [14–16]. Once identified, this zone represents an equilibrium point between outward forces originating from the tongue and inward forces originating from the lip and cheek musculature [14–16]. With mandibular denture teeth arranged to occupy this zone of muscular neutrality, the mandibular denture may achieve optimal physiologic and functional stability [14–16]. The technique for fabrication of neutral zone mandibular occlusal rim is described below.

Technique

1) The mandibular record base is positioned on the definitive cast.
2) The ridge area may be roughened with laboratory rotary instruments or a retention ledge may be incorporated in the design of the record base (Figure 4.7).

3) A bulk of modeling plastic impression compound can be preformed into a mandibular ridge shape by the technician/dentist. Molten black modeling plastic impression compound is applied along the crest of the mandibular record base.

4) The ridge surface of the previously formed modeling plastic impression compound rim is warmed and adapted to the record base.

5) The length of the anterior rim should correspond to the lip ruler measurement (esthetic space) taken during the diagnostic appointment (Figure 4.8a).

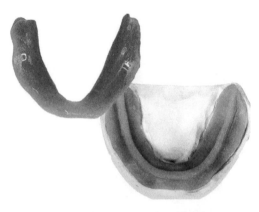

Figure 4.7 The ridge areas roughened with laboratory rotary instruments (left). Retention ledge incorporated in the design (right).

6) This measurement is transferred on the cast using the lip ruler and a compass (Figure 4.8(b)). The compass is set at an arbitrary point on the cast surface and the planned incisal length of the occlusal rim is evaluated (Figure 4.9a). It is rotated to the lateral surface of the cast and a line is scribed on the base of the cast to record the set measurement (Figure 4.9b).

7) The periphery of the modeling plastic rim is securely adapted to the record base with a heated instrument.

8) The compass is repositioned on the cast to evaluate the length of the anterior modeling plastic impression compound rim. A mark representing the predetermined length is scribed on the occlusal rim (Figure 4.10a). The rim is decreased or increased in length accordingly with a sharp blade (Figure 4.10b).

9) The height of the posterior modeling plastic impression compound rim length should correspond with the height of the retromolar pad.

10) The occlusal rim should be approximately 8–10 mm in width in the posterior region and 6–8 mm in width in the anterior region.

11) The rim is smoothened and further contoured in preparation for clinical use (Figure 4.11).

(a)

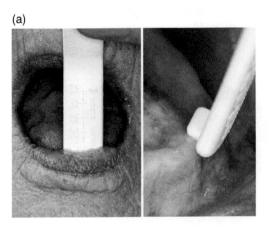

(b)

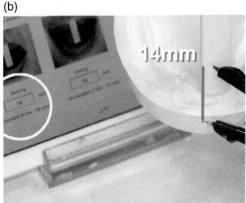

Figure 4.8 (a) The esthetic space is measured and recorded during the diagnostic appointment (left). The tab on the lip ruler is placed on the crest of the ridge while measuring esthetic space (right). (b) The lip ruler is appropriately positioned on mandibulary cast and the esthetic space measurement transferred using the compass.

(a)

(b)

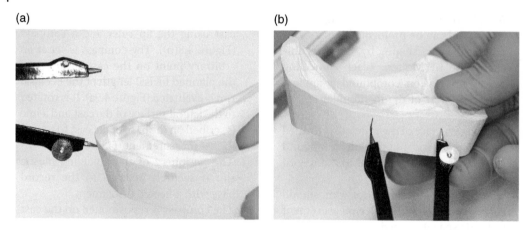

Figure 4.9 (a) The planned incisal length of mandibular occlusal rim. (b) The compass rotated laterally on the cast to scribe a line to record esthetic space measurement on cast.

(a)

(b)

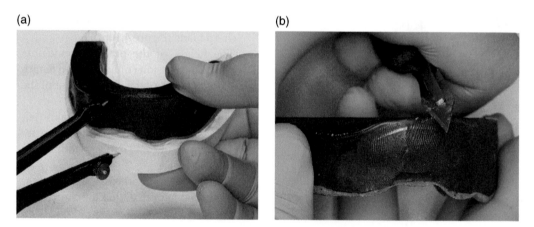

Figure 4.10 (a) The compass is repositioned on the cast and the predetermined length marked on rim. (b) The Occlusal rim is trimmed with a sharp blade to the predetermined length.

Figure 4.11 The Neutral zone mandibular occlusal rim occlusal view (left). The neutral zone mandibular occlusal rim lateral view (right).

Jaw Recorder Device (Central Bearing Device)

A simple, predictable, reproducible, and accurate method to locate and register the maxillomandibular relationship is by using a newly modified central bearing device called the jaw recorder device (designed by Massad). The new central bearing device permits physiologically unencumbered protrusive, retrusive, and lateral mandibular movements and permits their visualization as tracing marks on the striking plates [17]. It is indicated in the following clinical applications [17, 18]:

1) Recording occlusal vertical dimension (OVD).
2) Recording the CR for edentulous, partially edentulous, and implant patients.
3) Balancing the occlusion of definitive prostheses (complete dentures, partial dentures, and implant dentures).
4) Orthopedically repositioning the mandible.

Jaw Recorder Device Assembly

The jaw recorder device assembly consists of the following components (Figure 4.12):

- plastic striking plates (standard, small);
- pivoting nut and vertical threaded pin;
- centric pin receiver (available as small, large, and modified);
- clear plastic positioner disc.

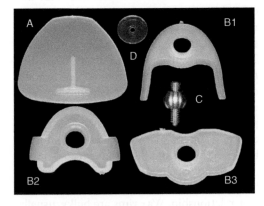

Figure 4.12 Jaw recorder device assembly:
A – striking plate; B1 – small pin receiver;
B2 – medium pin receiver; B3 – large pin receiver;
C – threaded pin; D – clear plastic positioner disc.

Mounting the Jaw Recorder Device on Record Bases

Record bases are fabricated for both the maxillary and the mandibular arches with a light polymerized resin material, as described previously. Record bases are designed to be slightly underextended, to minimize unwanted movement of the record bases during muscular / vestibular movements. The jaw recorder device stabilizes the jaws and aids in recording the maxillo-mandibular relationship at the predetermined OVD [17, 18]. The procedure of mounting the jaw-recorder device to the record bases is described below:

1) The underextended maxillary and the mandibular record bases are placed on their respective definitive casts.
2) A striking plate along with the large pin receiver (along with the pivoting nut and vertical pin) is utilized for the edentulous patient.
3) The striking plate is attached to the maxillary record base (Figure 4.13a). The procedure is described below:
   - A small portion of light polymerized resin is adapted to record base within confines of anterior palate. Resin should fill the anterior palatal vault up to the ridge crest. There should be enough resin material to secure the striking plate.
   - The striking plate is seated on the unpolymerized resin. The plate is positioned to extend beyond the periphery of the record base to ensure adequate contact area for the threaded vertical pin during excursive movements. It should be noted that the striking plate should be parallel to the residual ridge crest.
   - Resin polymerization is initiated with a curing light. In situations where a deep palatal vault is present the record bases may be directly placed in a light polymerization unit to initiate and complete the polymerization process.

(a)

(b)

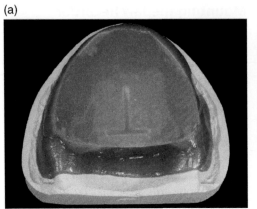
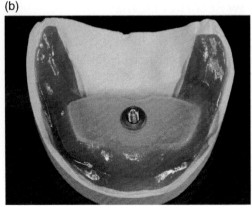

Figure 4.13 (a) Striking plate attached to maxillary record base; (b) pin receiver attached to mandibular record base.

4) A large pin receiver (with circular projection upward) is attached to mandibular record base (Figure 4.13b). The procedure for mounting the pin receiver on the mandibular record base is as follows:

- Two small ropes of light-polymerized resin are placed along the lingual flanges of the record base and extended up to the ridge crest.
- The pin receiver is seated on the unpolymerized resin on the lingual aspect of the mandibular record base such that the threaded pin will be located in the center of an imaginary line joining the second bicuspids.
- The pin receiver should be positioned such that it is parallel to the mandibular ridge crest. A second pin receiver may be utilized to facilitate the parallelism between the striking plate, pin receiver, and the residual ridges.
- Resin polymerization is initiated with a curing light.

Both record bases are placed in a light polymerization unit to complete the polymerization process. A very small amount of commercially available cyanoacrylate adhesive (such as Super Glue, produced by the original Super Glue Corporation, Ontario, Canada) is used to secure the light-activated resin and the striking plate assembly to the maxillary record base. Similarly, the same adhesive is used to secure the light-polymerized resin and the pin-receiver assembly to the mandibular record base. The pivoting nut along with the threaded pin is placed in the pin receiver. Upon completion of the mounting process, the security and adaptation of striking plate and pin receiver should be evaluated by applying inward and outward pressure on the assembly. If they become dislodged from their mountings, the mounting procedure is repeated as described above.

When the jaw recorder device is placed in the mouth, the dental practitioner may adjust vertical pin length to achieve the predetermined OVD. A common problem with the central bearing devices is that if they are not mounted correctly, the pin will not contact the striking plate in a perpendicular relationship. This will result in destabilization of the record bases and an inaccurate registration of the maxillo-mandibular relationship. The unique design of the new jaw recorder device permits repositioning of the pivoting nut to create a better perpendicular relationship between the pin and the striking plate.

Wax-only occlusal rims have commonly been used for recording the maxillo-mandibular relationship. Wax rims are bulky, usually unstable, and in the hands of an inexperienced practitioner may lead to registering of an inaccurate maxillo-mandibular relationship.

This would necessitate a remount procedure in the subsequent appointment [19–22]. Once the technique for using the jaw recorder device is adequately learned, it permits accurate and repeatable recording of the maxillo-mandibular relationship with minimal operator intervention and chair time [17, 18].

Mounting of the Jaw Recorder for Implant Overdentures

The procedure of mounting the jaw-recorder device on an arch to receive an implant-supported overdenture is similar to the one described previously (edentulous arch) in this chapter except that the record base should be fabricated with recesses to fit over the implant abutments (Figure 4.14). This

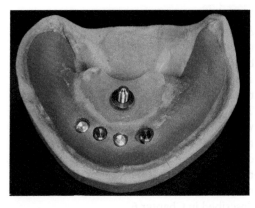

Figure 4.14 Recesses created in record base to fit over implant abutments.

prevents incomplete seating and rocking of the record base. After the record base is fabricated, the pin receiver or the striking plate can be attached in the same manner to the mandibular and maxillary arch respectively as discussed previously.

Mounting the Jaw Recorder Device on Partially Edentulous Arches

The jaw recorder device is also indicated for recording the maxillo-mandibular jaw relationship for partially edentulous patients. Its use aids the patient to perform mandibular movements at the established OVD without any tooth interference. It is also challenging to register the maxillo-mandibular relationship using wax occlusal rims for partially edentulous arches with mobile teeth as the natural teeth may move during the registration, resulting in an inaccurate recording. In these situations it may also be advantageous to use the jaw recording device as it permits the patient to perform mandibular movements at the established OVD without any tooth interference.

For partially edentulous arches, the record bases are fabricated with tissue stops in the edentulous area for improved stability (Figure 4.15a). Clasps can be incorporated in the record bases to aid in the retention of the record base (Figure 4.15b). The procedure for mounting the jaw recorder on partially edentulous

(a)

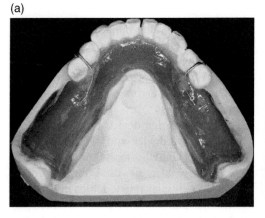

(b)

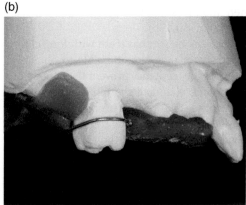

Figure 4.15 (a) Record base fabricated with tissue stops in edentulous area for improved stability; (b) retentive claps incorporated in record base for improved stability.

(a)
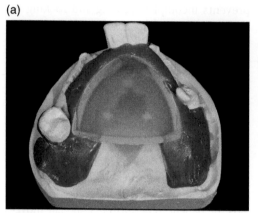

(b)
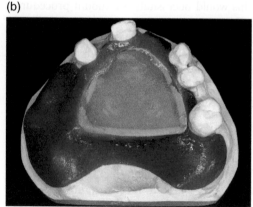

Figure 4.16 (a) Striking plate is attached to maxillary record base below occlusal plane for a partially edentulous arch; (b) striking plate adjusted and then attached to maxillary record base to avoid tooth impingement.

arches is similar to that described for edentulous arches except that the striking plate should be gently submerged in to the resin approximately 1 mm so that it lies below the occlusal plane (Figure 4.16(a)). If the striking plate impinges on the teeth it should be adjusted with slow-speed laboratory rotary instruments and then mounted on the maxillary record base (Figure 4.16b). The pin receiver should also be placed on the mandibular record base slightly below the occlusal plane (Figure 4.17a). A small pin receiver is indicated for partially edentulous arches. If the pin receiver impinges on the teeth it should be adjusted with slow-speed laboratory rotary instruments and then mounted on the mandibular record base (Figure 4.17b).

When the patient has teeth with long clinical crowns it may be impossible to obtain a record as the teeth may prevent the contact of the striking plate and the pin during record taking. The steep vertical overlap of the anterior teeth may also impede freedom of movement of the pin on the striking plate. Adjustment of the natural teeth is indicated when these situations occur to avoid extending the pin beyond the predetermined OVD. If these teeth are to be restored or extracted, they can be shortened before taking the recording by using a reduction coping as described below.

Procedure

A flexible vacuum / pressure-formed matrix is placed over the teeth (to be shortened) on the cast, and the matrix, as well as the teeth, are trimmed to the optimal height with laboratory rotary instruments (Figure 4.18a). This procedure helps to ensure that the teeth will not interfere with the centric relation records. The adjusted matrix (reduction coping) (Figure 4.18b) is then removed from the cast and seated on the natural teeth in the mouth. The tooth structure removal will be described in Chapter 6.

Summary

Record bases may be handmade, milled, or digitally printed. Regardless of the fabrication technique, it is critical that customized occlusal rims be fabricated to decrease the introduction of errors and increase the chances of success.

The jaw recorder device functions like a central bearing device and aids in accurate recording of maxillo-mandibular relationships. It may also be used for recording maxillo-mandibular relationships in partially edentulous patients and patients receiving implant prostheses. The jaw recorder device is disposable, except the vertical pin, which can be sterilized and reused.

(a)

(b)

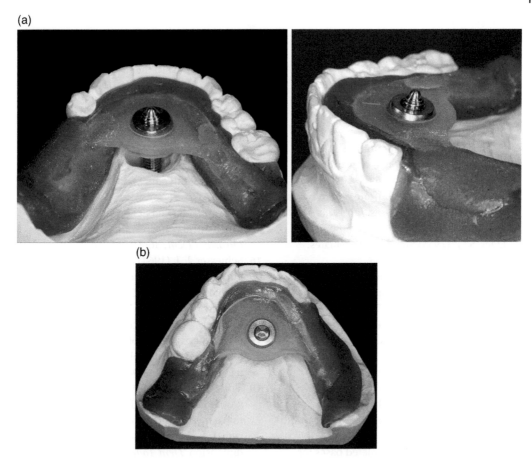

Figure 4.17 (a) Pin receiver attached to mandibular record base below occlusal plane, occlusal view (left), lateral view (right); (b) Small pin receiver adjusted and attached to mandibular record base to avoid tooth impingement.

(a)

(b)

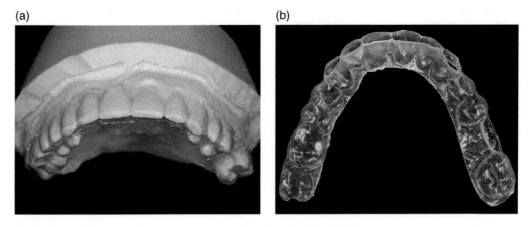

Figure 4.18 (a) Matrix and the teeth trimmed to the optimal height with laboratory rotary instruments; (b) adjusted matrix functions as a reduction coping.

References

1 Travaglini, E. A. (1962) Resilient tissue surface in complete dentures. *J Am Dent Assoc*, **64**, 512–517.

2 Klein, I. E., and Soni, A. (1979) Stabilized record bases for complete dentures. *J Prosthet Dent*, **42**, 584–587.

3 Ortman, H. R., and Edgerton, M. (1982) Utilizing the impression as a record base. *J Prosthet Dent*, **48**, 618–620.

4 Tucker, K. M. (1966) Accurate record bases for jaw relation procedures. *J Prosthet Dent*, **16**, 224–226.

5 Elder, S. T. (1955) Stabilized baseplates. *J Prosthet Dent*, **5**, 162–168.

6 McElroy, T. H., and Canada, K. W. (1984) Stabilized accurately fitting trial bases. *J Prosthet Dent*, **52**, 753–755.

7 McCracken, W. (1953) Auxiliary uses of cold curing acrylic resins in prosthetic dentistry. *JADA*, **47**, 298.

8 Alexander, J., and Beckett, H. (1997) A bilaminar record base for a complete overdenture with severe hard tissue undercuts: a clinical report. *J Prosthet Dent*, **77**, 233–234.

9 Graser, G. N. (1978) Completed bases for removable dentures. *J Prosthet Dent*, **39**, 232–236.

10 Helft, M., Cardash, H., and Kaufman, C. (1978) Combining final impressions with maxillomandibular relation records in stabilized record bases. *J Prosthet Dent*, **39**, 135–138.

11 Ortman, H. R., and Edgerton, M. (1982) Utilizing the impression as a trial denture base. *J Prosthet Dent*, **48**, 618–620.

12 Ortman, H. R. (1979) Relationship of the incisive papilla to the maxillary central incisors. *J Prosthet Dent*, **42**, 492–496.

13 Goldstein, R. E. (1998) *Esthetics in Dentistry. Biology of Esthetics*, 2nd edn. B. C. Decker Inc., Hamilton, ON, pp 101–123.

14 Fish, E. W. (1964) *Principles of Full Denture Prosthesis*, 6th edn, Staples Press, London, pp. 36–37.

15 Beresin, V. E., and Schiesser, F. J. (1976) The neutral zone in complete dentures. *J Prosthet Dent*, **36**(4), 356–367.

16 Schiesser, F. J. (1964) The neutral zone and polished surfaces in complete dentures. *J Prosthet Dent*, **14**, 854–865.

17 Massad, J. J., Connelly, M. E., Rudd, K. D., and Cagna, D. R. (2004) Occlusal device for diagnostic evaluation of maxillomandibular relationships in edentulous patients: A clinical technique. *J Prosthet Dent*, **91**, 586–590.

18 Wojdyla, S. M., and Wiederhold, D. M. (2005) Using intraoral Gothic arch tracing to balance full dentures and determine centric relation and occlusal vertical dimension. *Dent Today*, **24**, 74–77.

19 Phillips, G. B. (1927) Fundamentals in the mandibular movements in edentulous mouths. *J Am Dent Assoc*, **14**, 409.

20 Simpson, H. (1939) Registraion of centric relation in complete denture prosthesis. *J Am Dent Assoc*, **26**, 1682.

21 Schuyler, C. (1941) Intraoral method of establishing maxilla-mandibular relation. *J Am Dent Assoc*, **28**, 17.

22 Gysi, A. (1910) The problem of articulation. *Dent Cosmos*, **52**, 1.

5

Developing an Esthetic Blueprint

Introduction

Complete dentures must be attractive to patients, and to their family members and friends. The esthetic surface of the denture is the surface where the anterior prosthetic teeth are placed and positioned. The anterior prosthetic teeth may be oriented and positioned in the labial, palatal / lingual and mesial / distal axes to develop a pleasant, esthetic, and natural smile for the patient. This chapter focuses on the development of the esthetic surface in conformity with a natural feel to develop an esthetically acceptable prosthesis for the patient. Please note that the procedures accomplished in Chapters 5, 6 and 7 are all carried out at the same appointment; they have been compartmentalized as separate chapters for easy understanding.

Contouring and Shaping the Maxillary Occlusal Rim

The wax occlusal rim should be developed so that it is in harmony with the facial form. It should be contoured to fill the space occupied by the patient's natural teeth [1]. Prosthetic teeth should be set in this space, and not on the crest of the resorbed ridge, to achieve optimal esthetics, soft tissue support, and function [1]. The maxilla resorbs centripetally (upward, medially, and rearward) (Figure 5.1), so positioning the prosthetic teeth on the

crest of the ridge in these patients places the teeth lingual to the position previously occupied by natural teeth. This may lead to poor esthetics (due to inadequate support of the facial tissues) and compromised function [1, 2].

A clinically adjusted and well contoured maxillary wax occlusal rim may be termed an "esthetic blueprint" because it serves as a guide for positioning the prosthetic teeth based on patient esthetics and phonetics. The esthetic blueprint conveys information about the midline, high smile line, labial inclination for lip support, tooth display in repose and during smiling, as well as the orientation of the occlusal plane. This information is then transferred to the prosthetic technician (through the esthetic blueprint) for the development of the wax trial denture. A maxillary record base with the wax occlusal rim is fabricated prior to this clinical appointment as per the patient specifications described in Chapter 4. The procedure of adjusting and contouring the maxillary wax occlusal rim to develop the esthetic blueprint is described below:

1) The intaglio surface and the flanges of the record base are inspected for any sharp spots or rough surfaces and adjusted as needed.
2) The maxillary record base with the wax occlusal rim is placed in the patient's mouth. The retention and stability of maxillary record base should be evaluated. If the maxillary record base is loose,

Application of the Neutral Zone in Prosthodontics, First Edition. Joseph J. Massad, David R. Cagna, Charles J. Goodacre, Russell A. Wicks and Swati A. Ahuja.
© 2017 John Wiley & Sons, Inc. Published 2017 by John Wiley & Sons, Inc.
Companion website: www.wiley.com/go/massad/neutral

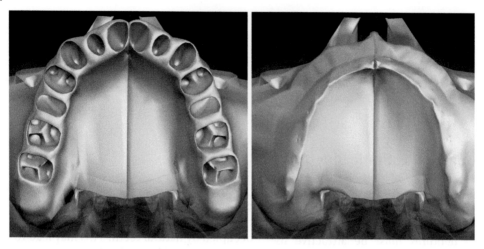

Figure 5.1 Animated depiction of the maxillary ridge following extraction of teeth (left) and the change in maxillary ridge form following ridge resorption (right).

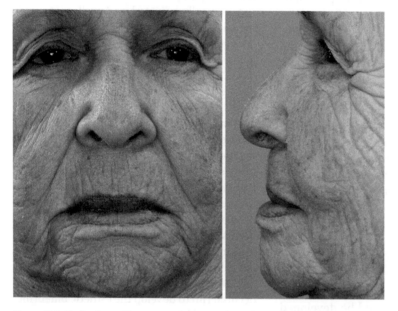

Figure 5.2 Evaluation of lip support in repose: frontal view (left); profile (right).

its retention may be improved by using a small amount of denture adhesive. In this situation, it should be explained to the patient that the record base is not the actual maxillary denture and the definitive denture will have superior retention to that of the record base.

3) The maxillary wax occlusal rim should be optimally contoured to provide adequate support to the upper lip [3, 4]. Lip support

is evaluated by asking the patient to approximate the lips and visually inspecting the patient from the frontal aspect and in profile (Figure 5.2). A well-supported upper lip forms a right angle with the base of the nose or even slightly acute angle with the base of the nose. The age and personal characteristics of the individual patient must also be considered while evaluating the lip support. If the lip support appears inadequate, a strip of baseplate wax can be added to the labial aspect of the wax rim and contoured as necessary. If the lip support is excessive, a heated rim former can be used to contour the wax rim to the desired thickness and inclination. An optimally contoured wax rim guides the antero-posterior positioning of the prosthetic teeth and supports the lip in a natural and pleasing manner.

4) The incisal length of the wax rim is determined using esthetics and phonetics as a guide. The incisal length of the maxillary wax occlusal rim is evaluated by checking the height of the wax rim compared with the patient's relaxed upper lip [5–9]. The incisal length of the wax rim will guide the placement of the incisal edges of the maxillary anterior teeth. Initially, the wax rim is maintained at the same height or slightly longer than the relaxed upper lip length when the lips are parted (Figure 5.3). The teeth can also be positioned cervically or incisally depending on the age and sex of the patient [4]. The edge of the wax rim (and eventual incisal edges of the prosthetic teeth) should touch the wet-dry line of the lower lip while making the "F" and "V" sounds (Figure 5.4) [6].

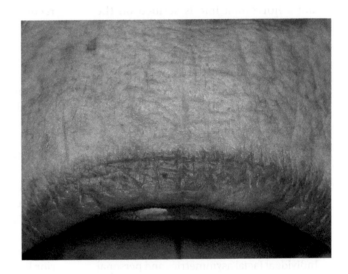

Figure 5.3 Wax rim is 1–2 mm longer than the relaxed upper lip when the lips are separated.

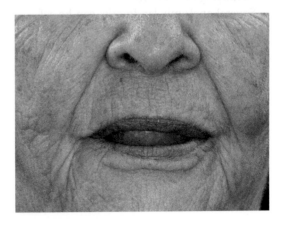

Figure 5.4 Rim touching the wet dry line of the lower lip when executing fricative ("F" and "V") sounds.

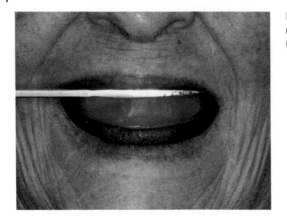

Figure 5.5 A horizontal line is scribed on the wax rim to record maximum elevation of the upper lip (high smile line).

5) The patient is asked to smile maximally and a horizontal line is scribed on the wax rim to record maximum elevation of the upper lip (high smile line) (Figure 5.5). This measurement helps to guide the practitioner in selecting the length and cervical position of the neck of the maxillary anterior teeth for controlled esthetics [4]. Generally, 75–100% of the length of the maxillary anterior teeth should be visible in a patient's smile [3].

6) A straight wax knife is used as a guide to locate the facial midline and the position and orientation of the midline is transferred to the wax rim by scoring a vertical line in the rim (Figure 5.6) [10, 11]. Individual facial asymmetries and personal characteristics should also be factored into identifying this line [10].

7) Several anatomic landmarks can be referenced to assess the potential dimensions of the prosthetic teeth including : bizygomatic width [12], intercommissural distance [13], interpupillary distance [13], width of the philtrum [14], interalar width [15–17], and intercantral width [18, 19]. It has been recommended that a combination of such factors should best be used to determine tooth selection and placement [20]. In this technique a sliding caliper is adjusted to approximate the interalar width (Figure 5.7a) and the same markings and their positions are recorded on the wax rim (Figure 5.7b). On average these marks correspond to the intercuspid distance or combined width of the anterior teeth [15–17].

8) Anatomical landmarks and esthetic presentation have been considered to determine the orientation of the occlusal plane [1] for the edentulous maxilla. A Fox plane [1] or a tongue blade can be used to evaluate the parallelism between the anterior plane of the wax rim and the interpupillary line as well as the posterior plane of the wax rim and the ala-tragus line (Figure 5.8) [21–23]. The ala-tragus line is initially used as a reference for establishing the orientation of the posterior length of the rim [21–23]. The posterior plane may need to be relocated later in respect of the deferential location of the mandibular plane. A heated rim former can be used to alter and adjust the wax rim lengths so that the wax rim is parallel to the anatomical determinants. The orientation of the occlusal plane should also consider the smile line of the lower lip [24].

9) Finally, the accuracy of rim length (in repose and during full animation), contour, anteroposterior plane and all the markings are verified by repositioning wax rim in the mouth and evaluating the combined factors.

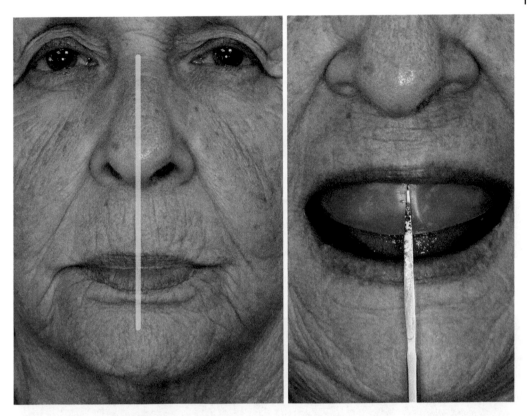

Figure 5.6 Facial midline determination.

(a)　　　　　(b)

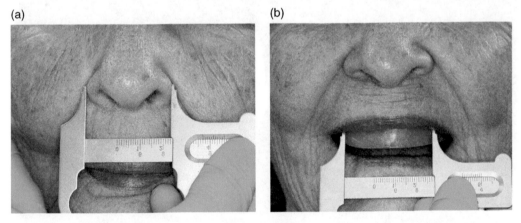

Figure 5.7 (a) Sliding caliper adjusted to approximate the interalar width, and (b) the distance transferred to the wax rim.

As an additional technique, facial veneer shell teeth can be used to preview the esthetic positioning of the teeth and can also be waxed into the occlusal rims and utilized for evaluating esthetics and phonetics during this appointment (Figure 5.9). These are available in several sizes and should be selected based on the established parameters in the wax rim. The incisal edge position of the shell teeth should correspond to predetermined length of

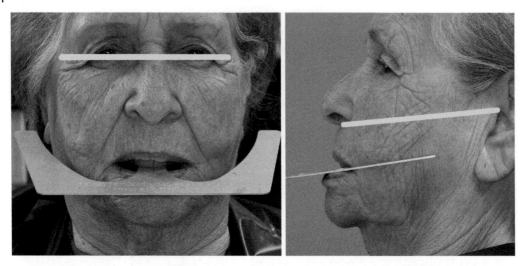

Figure 5.8 Location of the Occlusal plane based on the interpupillary line (left). Location of the Occlusal plane based on ala-tragus line (right).

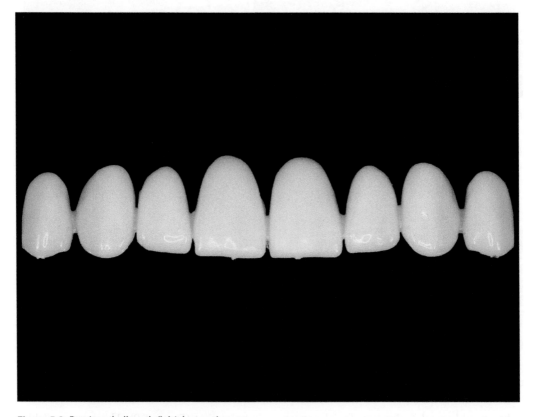

Figure 5.9 Preview shell teeth (labial aspect).

the wax rim. When waxing the shell teeth on the occlusal rims, the midline marked on the wax rim should be extended to the occlusal surface of the wax occlusal rim so that it can serve as a reference later in the procedure (Figure 5.10). The width and the labial inclination of the wax occlusal rim should be reduced appropriately to permit optimal

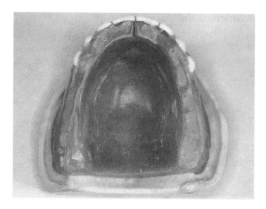

Figure 5.10 Midline extended to the incisal/occlusal surface of the wax rim and adaptation of shell teeth.

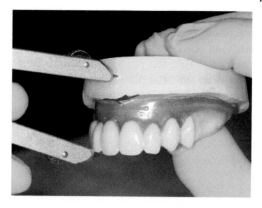

Figure 5.11 Pre-determined incisal length of the wax rim guides the incisal edge positioning of shell teeth.

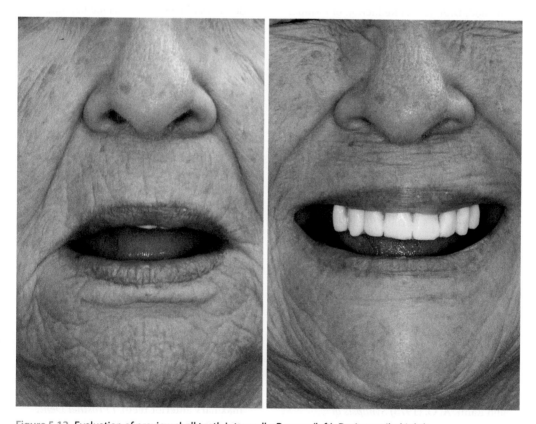

Figure 5.12 Evaluation of preview shell teeth intraorally. Repose (left). During smile (right).

labio-lingual positioning and incisal positioning, as well as inclination of the shell teeth (Figure 5.11).

This procedure permits patient viewing of the shell teeth to approve or disapprove the established incisal edge position during both repose and smiling (Figure 5.12). It also helps to acquire the patient's input for the established midline, high smile line, the esthetic display, general size and shape, and any noticeable asymmetry before the actual trial placement appointment. The patient's comments at this time can be addressed by the practitioner and if possible alterations can be made at this time. Keeping the patient involved during the process can eliminate a surprise at the trial placement appointment and patient disappointment, requiring resetting of the prosthetic teeth and another try-in appointment. Use of the facial veneer teeth helps verify tooth positioning and prevents any confusion in the minds of the laboratory technician, the patient, and the practitioner related to markings on the pink baseplate wax.

Summary

The maxillary occlusal rim should be contoured appropriately based on patient esthetics and phonetics and anatomic structures, including the muscles at rest and in function.

References

1 Engelmeier, R. (1996) Complete denture esthetics. *Dent Clin N Am*, **40**, 71–84.

2 Atwood, D. A. (1971) Reduction of residual ridges: a major oral disease entity. *J Prosthet Dent*, **26**, 266–279.

3 Hock, D. A. (1992) Qualitative and quantitative guides to the selection and arrangement of the maxillary anterior teeth. *J Prosthodont*, **1**, 106–111.

4 Frush, J. P., and Fisher, R. D. (1958) The dynesthetic interpretation of the dentogenic concept. *J Prosthet Dent*, **8**, 558–581.

5 Boucher, C. O., Hickey, J. C., and Zarb, G. A. (1975) *Prosthodontic Treatment for Edentulous Patients*. 7th edn. Mosby, St Louis, MO.

6 Plummer, K. D. (2009) Maxillomandibular records and articulators, in *Textbook of Complete Dentures*, 6th edn (eds. A. O. Rahn, J. R. Ivanhoe, and K. D. Plummer). People's Medical Publishing House, Shelton, CT, pp. 161–181.

7 Sharry, J. J. (1974) *Complete Denture Prosthodontics*, 3rd edn. McGraw-Hill, New York, NY.

8 Ellinger, C. W., Ravson, J. H., Terry, J. M., and Rahn, A. O. (1975) *Synopsis of Complete Dentures*, Lea & Febiger, Philadelphia, PA.

9 Vig, R. G., and Brundo, G. C. (1978) The kinetics of anterior tooth display. *J Prosthet Dent*, **39**, 502–504.

10 Beyer, J. W., and Lindauer, S. J. (1998) Evaluation of dental midline position. *Semin Orthod*, **4**, 146–152.

11 Miller, E. L., Bodden, W. R. Jr, and Jamison, H. C. (1979) A study of the relationship of the dental midline to the facial median line. *J Prosthet Dent*, **41**, 657–660.

12 Gueye, M., Dieng, L., Mbodj, E. B., *et al.* (2014) Relationship between bizygomatic width and the size of maxillary anterior teeth among young Senegalese black people recruited in army. *Odontostomatol Trop*, **37**, 5–12.

13 Kini, A. Y., and Angadi, G. S. (2013) Biometric ratio in estimating widths of maxillary anterior teeth derived after correlating anthropometric measurements with dental measurements. *Gerodontol*, **30**, 105–111.

14 Vuttiparum, N., and Benjakul, C. (1989) Relationship between the width of maxillary central incisors and philtrum. *J Dent Assoc Thai*, **39**, 233–239.

15 Hoffman, W., Bomberg, T. H., and Hatch, R. A. (1986) Interalar width as a guide in denture tooth selection. *J Prosthet Dent*, **55**, 219–221.

16 Mavroskoufis, F., and Ritchie, G. M. (1981) Nasal width and incisive papilla as guides for the selection and arrangement of maxillary anterior teeth. *J Prosthet Dent*, **45**, 592–597.

17 Wehner, P. J., Hickey, J. C., and Boucher, C. O. (1967) Selection of artificial teeth. *J Prosthet Dent*, **18**, 222–232.

18 Abdullah, M. A. (2002) Inner canthal distance and geometric progression as a predictor of maxillary central incisor width. *J Prosthet Dent*, **88**, 16–20.

19 Al Wazzan, K. A. (2001) The relationship between intercanthal dimension and the widths of maxillary anterior teeth. *J Prosthet Dent*, **86**, 608–612.

20 Scandrett, F. R., Kerber, P. E., and Umrigar, Z. R. (1982) A clinical evaluation of techniques to determine the combined width of the maxillary anterior teeth and the maxillary central incisor. *J Prosthet Dent*, **48**, 15–22.

21 van Niekerk, F. W., Miller, V. J., and Bibby, R. E. (1985) The ala-tragus line in complete denture prosthodontics. *J Prosthet Dent*, **53**, 67–69.

22 Chrystie, J. A. (1985) A device for establishing the occlusal plane for complete dentures. *J Prosthet Dent*, **54**, 447.

23 Husseinovitch, I., and Chidiac, J. J. (2002) A modified occlusal plane device. *J Prosthet Dent*, **87**, 240.

24 Camara, C. A. (2010) Aesthetics in orthodontics: Six horizontal smile lines. *Dental Press J Orthod*, **15**, 118–131.

6

Registering the Maxillo-Mandibular Jaw Relationship

Introduction

Conventional complete denture therapy has a rich tradition of innovative techniques, clever processes, unique materials, and clinical precision [1]. Although the principles have almost remained the same over the years, modern complete denture wearers can expect to benefit from subtle revisions in conventional treatment philosophies [1]. Use of a classic concept incorporating a traditional device to assist in establishing and registering centric relation is applicable to today's removable prosthodontic patients [1]. This chapter describes the procedures for registering maxillo-mandibular jaw relationships using a modified central bearing device. Record bases and occlusal rims are used to make the centric relationship records and transfer the correct orientation of the casts to the articulator. It is important to select an appropriate articulator prior to this appointment [2]. A Class IV (fully adjustable) articulator is not required for denture fabrication [2]. A Class III articulator (semiadjustable) (Figure 6.1) is suitable to develop a proper occlusion for complete dentures [2, 3]. These articulators accept facebow and interocclusal records. Some also accept protrusive and lateral records, which may be registered during the jaw relation recording, or the trial placement appointment.

Facebow Recording

Mounting the maxillary cast on the articulator relative to the transverse horizontal axis (often described by the term hinge axis) of the patient is an important consideration in restorative dentistry [4]. Arbitrarily mounting the maxillary cast results in a discrepancy between the arc of closure of the articulator and the arc of closure of the patient. In such situations it is impossible to verify the position of the mandibular cast relative to the maxillary cast using interocclusal records made at increased vertical dimension of occlusion or increase the occlusal vertical dimension (OVD) on the articulator [4]. An occlusion developed on an articulator with an inaccurate arc of closure will also have several potential interferences and deflective contacts. To avoid these problems, it is recommended that a facebow record be used to orient the maxillary cast correctly to the condylar elements of the articulator [4–6].

There are two types of facebows: kinematic and arbitrary [6]. The kinematic facebow precisely relates the maxillary cast to the true hinge axis of the patient [6]. The arbitrary facebow relates the maxillary cast to within 5 mm anatomic average of the hinge axis of the patient. An arbitrary facebow is recommended to be acceptable for complete denture fabrication [6]. The procedure for making a facebow record with the quick

Application of the Neutral Zone in Prosthodontics, First Edition. Joseph J. Massad, David R. Cagna, Charles J. Goodacre, Russell A. Wicks and Swati A. Ahuja.
© 2017 John Wiley & Sons, Inc. Published 2017 by John Wiley & Sons, Inc.
Companion website: www.wiley.com/go/massad/neutral

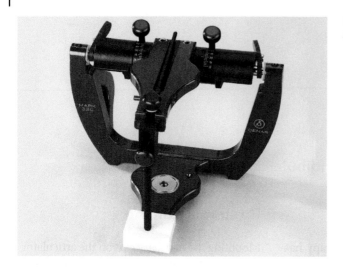

Figure 6.1 Class III semi adjustable articulator.

mount facebow (Denar Slidematic Facebow, Whipmix) is described below:

1) Notches are created with a sharp blade on the posterior aspects of the clinically adjusted maxillary wax occlusal rim (esthetic blue print) for indexation.
2) The clinically adjusted maxillary wax occlusal rim is placed in the mouth. If the patient has a mandibular denture it is also placed in the mouth as it can be used to support the facebow fork while making the facebow record.

   The facebow fork is loaded with an interocclusal registration material (vinyl polysiloxane, baseplate wax, or modeling plastic impression compound) and placed over the maxillary wax rim such that the center of the fork is aligned with the center of the wax rim (Figure 6.2). An alternative method can be used in which the facebow fork is centered and attached to the rim outside the mouth.
3) Upon polymerization of the registration material, the facebow fork is removed from the oral cavity and the excess is trimmed with a sharp blade.
4) With the patient sitting upright and looking straight forward, the bite fork assembly is replaced in the mouth. The assembly is stabilized by asking the patient to occlude on cotton rolls.

5) Next, the facebow frame (caliper) is positioned by inserting the ear pieces and tightening the caliper thumb screw.
6) The anterior point of reference (43 mm above the incisal edge of maxillary right lateral incisor) is located using the reference plane locator (included in the assembly) and marked with an indelible marker. The reference pointer on the facebow is aligned with the anterior point of reference. The properly aligned facebow assembly is secured and made rigid by tightening the remaining adjustment screws (Figure 6.3).
7) The caliper thumb screw is loosened and the ear pieces are removed. The patient is instructed to open and the record base is gently dislodged from the maxilla. The facebow record is transported out of the mouth in a forward direction. The record can now be transferred to the articulator (Figure 6.4) and maxillary master cast mounted, relative to the hinge axis (Figure 6.5).

Centric Relation Recording for the Edentulous Patient

A primary goal of restorative dentistry is to establish a harmonious relationship between the restoration, bone, temporo-mandibular

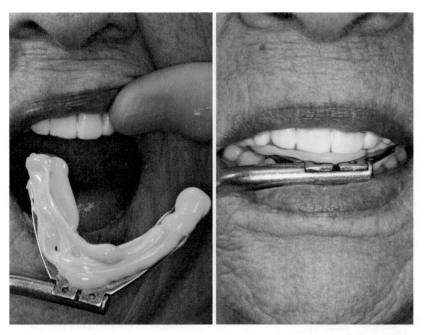

Figure 6.2 The facebow fork loaded with interocclusal registration material and inserted in the mouth (left). Center of fork aligned with the midline (right).

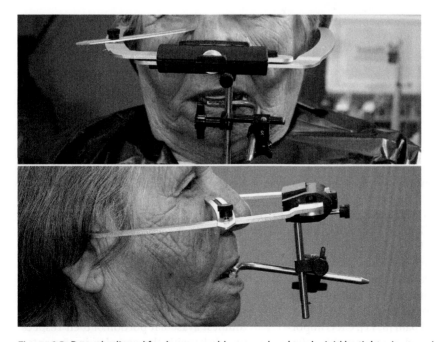

Figure 6.3 Properly aligned facebow assembly secured and made rigid by tightening remaining adjustment screws. Frontal view (top); Profile (bottom).

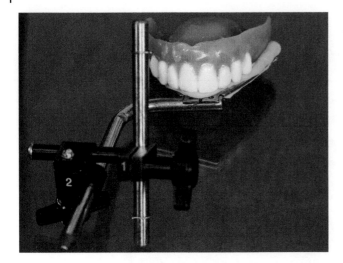

Figure 6.4 Quick mount facebow record assembly.

Figure 6.5 Facebow assembly mounted on the articulator and the maxillary cast carefully joined to the record.

joints, muscles, and ligaments [7]. This is best accomplished by restoring a patient in the centric relation (CR) position at the pre-established OVD [7, 8]. CR is stable, reproducible, and a reference position [8–11]. Various studies have indicated that the CR position is the ideal treatment position for procedures involving fixed prosthodontics, removable prosthodontics, orthodontics, and TMD and facial pain [8, 11]. CR is important because it is used as a repeatable reference position for restoration of the occlusion.

Errors in recording this position may have detrimental clinical implications. Accurately recording CR and establishing the occlusion coincident with this position is a critical factor in restorative dentistry [12]. CR is defined as:

the maxillo-mandibular relationship in which the condyles articulate with the thinnest avascular portion of their respective disks with the complex in the anterior-superior position against the slopes of the articular eminencies. This position is independent of tooth contact. This position is clinically discernible when the mandible is directed superiorly and anteriorly. It is restricted to a purely rotary movement of the mandible about the transverse horizontal axis. [13]

Most authors agree that the CR record is an important but difficult maxillo-mandibular relationships to register accurately [14, 15]. The ability to record this relationship reliably depends on various factors including a detailed understanding of the anatomy and physiology of temporomandibular joint (TMJ), experience and skill of the practitioner, proper handling of materials and devices, and cooperation of the patient. Several techniques for recording the CR position have been described in the literature, including direct check bite interocclusal records, graphic records (intraoral and extraoral), and functional records [16, 17].

The graphic recording method was popular in the early 1900s. It was modified and revised by several clinicians and researchers and has recently been updated [18–20]. A modern central bearing device permits simultaneous registration of horizontal and vertical jaw relationships with a single record [20]. The principal value of the central bearing device is that it eliminates physical interferences (of teeth and/or occlusal rims). In addition, an external force or operator manipulation is not required to make a graphic record. Advantages of the modified central bearing device (jaw-recorder device) include: ease of attachment to record bases using conventional light-activated resin material and a unique pivoting central bearing pin for adjustment in all dimensions during the recording process. All components are disposable, and mounting the device on the record bases is quick and easy [20].

Before recording CR, it is crucial to first establish the patient's OVD [20]. Most patients with existing dentures have diminished OVD [20]. To improve esthetics and function, and permit optimal recording of vertical and horizontal maxillo-mandibular relationships in these patients, OVD must first be reestablished [20]. As discussed in Chapter 2, an orthopedic splint can be fabricated for the patient to evaluate proposed alterations of existing OVD and for orthopedic resolution of mandibular posture, prior to fabrication of new prosthesis. It becomes extremely easy to make records when the patient is preconditioned to an acceptable OVD, muscles are relaxed, and the condyles appropriately seated in the glenoid fossa. The technique for registering the CR record is described below [20].

Maxillary and mandibular record bases with the mounted jaw recorder device are fabricated prior to this clinical appointment as discussed in Chapter 4 (Figure 6.6). Intaglio surface and flanges of the record base are inspected for sharp or rough areas and adjusted as needed [2].

1) Using a small indelible marker, a small dot is placed on the patient's nose and another on the patient's chin.
2) The orthopedic occlusal device previously fabricated for the patient is placed in the mouth (described in detail in Chapter 2), and the patient is asked to close gently until occlusal contact is achieved and the vertical distance between the dots is measured using a caliper or ruler. This linear distance represents the treatment/desired OVD (Figure 6.7).
3) The orthopedic occlusal device is removed from the mouth and the maxillary and mandibular record bases with mounted jaw recorder device are placed in the mouth. The patient is asked to close gently until the threaded pin contacts the striking plate and the vertical distance between the dots is measured using a caliper or ruler.
4) The difference between the treatment OVD and the OVD with the record bases is noted.

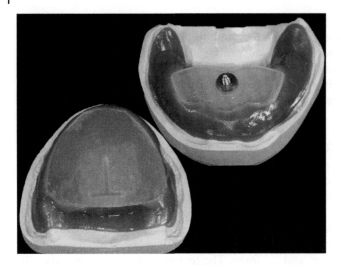

Figure 6.6 Maxillary and mandibular record bases with mounted jaw recorder device.

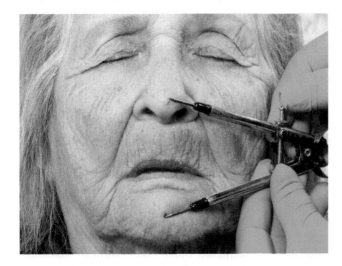

Figure 6.7 Treatment OVD established with orthopedic occlusal device being registered.

5) The threaded pin can be adjusted as necessary until contact with the striking plate reestablishes the treatment OVD (Figure 6.8). If the patient does not have existing dentures, the treatment OVD can be established for the patient at this appointment as discussed in Chapter 2. The pin position should also be verified to ensure that the pin contacts the striking plate in a perpendicular relationship. The unique design of this jaw recorder device permits repositioning of the pin to create better perpendicularity with the striking plate.

6) The maxillary record base is removed from the mouth and the striking plate is painted using a permanent indelible ink marker (Figure 6.9).

7) The maxillary record base is replaced in the mouth and the patient is asked to re-establish contact between the threaded pin and striking plate. The patient is also instructed to maintain this contact throughout the registration procedure (Figure 6.10a).

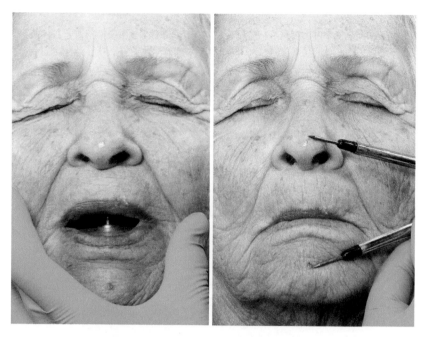

Figure 6.8 OVD greater than the treatment OVD (left). Threaded pin adjusted to restore the OVD to treatment OVD (right).

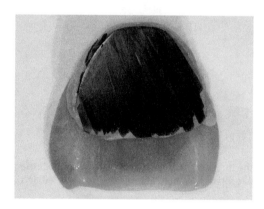

Figure 6.9 Striking plate painted using permanent indelible ink marker.

8) Next, the patient is encouraged to move the mandible forward (Figure 6.10b) and backward several times. Originating at the posterior-most mandibular position, the patient is instructed to make left and right lateral mandibular movements, always returning to the posterior most position (Figure 6.10c). Cumulatively, these mandibular movements will generate a Gothic arch tracing (arrow point) on the striking plate. A clear and distinct arrow point should be generated (Figure 6.11). The procedure is repeated until a discernible arrow is captured. The apex of the arrow is taken to represent the CR position of the mandible.

9) The maxillary record base is removed from the mouth and a clear plastic receiver disc is attached to the striking plate (Figure 6.12). The hole of the receiver disc is positioned to correspond with the tip of the arrow point tracing. Sticky wax is used to attach the receiver disc to the striking plate.

10) The maxillary record base is replaced in the mouth and the patient is instructed to close so that the threaded pin approximates the hole of the receiver disc.

11) With the mandible closed in this position, occlusal registration material is injected between the maxillary and mandibular record bases to register the CR position of the mandible at the established OVD (Figure 6.13a and b).

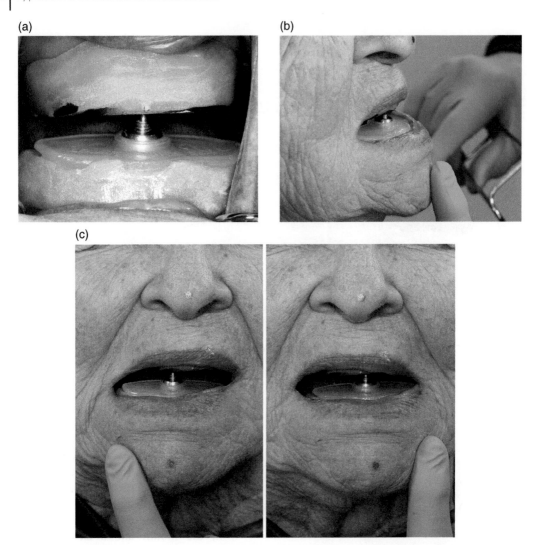

Figure 6.10 (a) Patient asked to maintain contact between the threaded pin and striking plate throughout the tracing procedure; (b) patient instructed to move the mandible forward; (c) patient instructed to make left and right lateral mandibular movements.

12) After the polymerization of the registration material, the maxillary and mandibular record bases are carefully removed from the mouth (Figure 6.13c).

Lateral and/or protrusive interocclusal records can also be registered, to program the articulator if a balanced occlusal scheme is planned for the patient [21]. It is important to observe that there are no inaccuracies of the record bases when making intraoral records. If an interference exists, it should be relieved. If the wax has dislodged from the base, it should be re-attached by heating (Figure 6.13d). In the presence of such confounding errors the process of making a CR record should be repeated.

The centric relation record and bases are reassembled and placed on the maxillary cast with the articulator inverted. The mandibular cast is carefully placed into the record and can now be mounted in the articulator (Figure 6.13e).

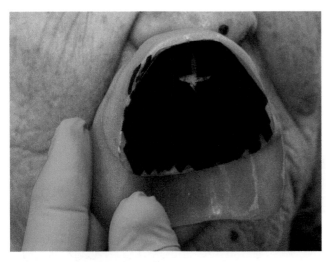

Figure 6.11 Gothic arch tracing (arrow point) on the striking plate. Note: the apex of the arrow is taken to represent CR position of the mandible.

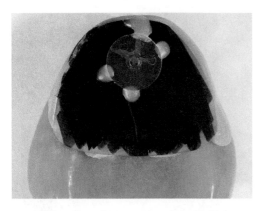

Figure 6.12 Clear plastic receiver disc attached to striking plate. Note: the hole of the receiver disc is positioned to correspond with the tip of the arrow point tracing.

Centric Relation Recording for the Partially Dentate Patient

The jaw recording device can also be adapted for use with patients who have remaining dentition and require restorative rehabilitation. The occlusal relationships must be considered in the design and location of the maxillary and mandibular record bases. Tooth mobility and record base stability must also be assessed. The record bases can be developed with the striking plate and opposing threaded pin to approximate the existing occlusal plane (Figure 6.14a). With the record bases in place and the pin adjusted to the desired OVD, the teeth are not in contact (Figure 6.14b). Once eccentric jaw movements are initiated, any occlusal contacts encountered should be disclosed with articulating paper. These intraoral marks can be visualized and transferred to the master cast (Figure 6.14c). A clear pressure formed thermoplastic material (Clear splint Biocryl 1.5 mm/125 mm, Great Lakes Orthodontics, Tonawanda, NY) is used to create a matrix on the dental arch that contains the identified tooth contacts/interferences. Rotary instrumentation is used to reduce the areas of contact through the matrix. The matrix is removed from the master cast and trimmed to cover adjacent teeth. This becomes the reduction guide, which is placed intraorally (Figure 6.14d). By carefully using this guide, tooth structure can be removed to emulate the same reduction on the master cast. This procedure is repeated until no further contacts/interferences are identified during eccentric jaw movements (Figure 6.14e). The same procedure may be performed on the opposing arch to gain adequate occlusal clearance if required. Without

(a)

(b)

(c)

(d)

(e)

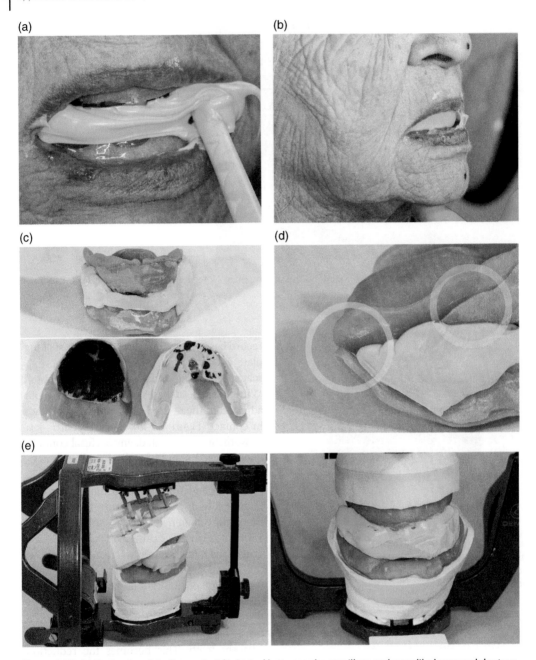

Figure 6.13 (a) Occlusal registration material injected between the maxillary and mandibular record denture bases to register CR position of mandible at established OVD. (b) The CR record being registered at the predetermined OVD. (c) Registration material, the maxillary and mandibular record bases are removed from the mouth (top). Maxillary and mandibular record bases are carefully separated from each other (bottom). (d) Inaccuracies present including posterior interfering contact of the record bases and separation of the wax rim from the maxillary record base. (e) The centric relation record reassembled and carefully placed on the maxillary cast with the articulator inverted. The mandibular cast is carefully placed into the record (left). Mandibular cast mounted in the articulator (right).

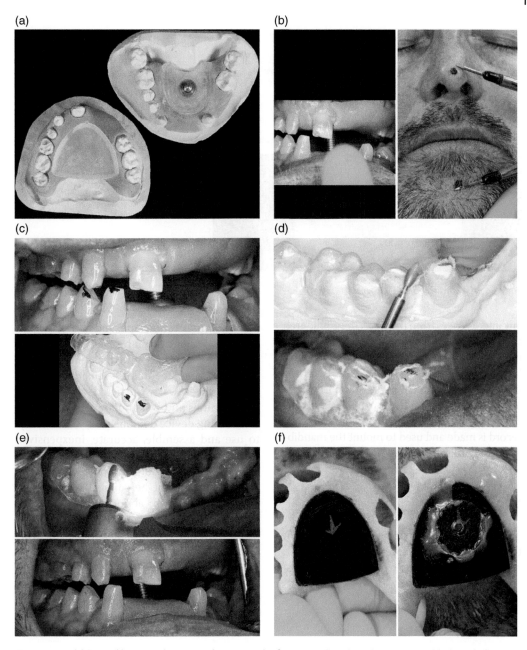

Figure 6.14 (a) Record bases with mounted jaw recorder for a partially edentulous patient. (b) Threaded pin adjusted to desired OVD (left) as recorded by points on nose and chin (right). (c) Occlusal contacts identified with articulating paper (top) and transferred to cast (bottom). (d) Matxix placed on the cast and the marked contacts eliminated (top). Matrix placed in the mouth to guide tooth reduction (bottom). (e) Contacts/ interferences adjusted (top) until eliminated (bottom). (f) Gothic arch tracing (left). Pin receiver attached over apex of the tracing (right). (g) CR registered at established OVD. Patient postured in CR position (top, left). Occlusal registration material is injected between the maxillary and mandibular record bases to register CR (top, right). Casts are mounted in articulator (bottom).

(g)

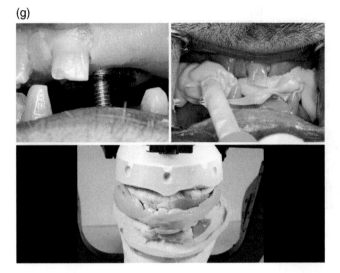

Figure 6.14 (Continued)

any interferences from the teeth, the gothic arch tracing is easy to establish as the jaw can now move to any eccentric position while maintaining pin contact on the striking plate. As previously described, the CR jaw position is located at the arch apex with the pin centered on the receiver (Figure 6.14f), and an interocclusal record is made and used to mount the mandibular cast in the articulator (Figure 6.14g).

Summary

The central bearing device is presented in this chapter. This intraoral gothic arch tracer, records the mandibular movements in one plane and registers an apex indicating the CR position. Other central bearing devices used in the past were difficult to assemble, cumbersome, and technique sensitive; hence these devices were restricted to academic communities and prosthodontists. The central bearing device demonstrated here is easy to use and assemble, accurate, inexpensive, and reliable.

Although applicable in most cases, this technique may not be suitable for patients who present with abnormal occlusal or skeletal relationships, steep vertical overlap of the anterior teeth, or insufficient neuromuscular competency required to make instructed jaw movements.

References

1 Massad, J. J., Cagna, D. R., Lobel, W. A., and Thornton, J. P. (2008) Complete denture prosthodontics: Modern approaches to old concerns. *Inside Dentistry*, **4**(8), 84–93.
2 Plummer, K. D. (2009) Maxillomandibular records and articulators, in *Textbook of Complete Dentures*, 6th edn. (eds. A. O. Rahn, J. R. Ivanhoe, K. D. Plummer). People's Medical Publishing House, Shelton, CT, pp. 161–181.
3 Márton, K., Jáhn, M., and Kivovics, P. (2000) The use of the Dentatus articulator in complete denture prosthetics. *Fogorv Sz*, **93**(1), 23–28.
4 Wilkie, N. D. (1979) The anterior point of reference. *J Prosthet Dent*, **41**, 488–496.
5 Thorp, E. R., Smith, D. E., and Nicholls, J. I. (1978) Evaluation of the use of a face-bow in complete denture occlusion. *J Prosthet Dent*, **39**(1), 5–15.

6 Schallhorn, R. G. (1957) A study of the arbitrary center and kinematic center of rotation for face bow mountings. *J Prosthet Dent*, **7**, 162–169

7 Atwood, D. A. (1968) A critique of research of the posterior limit of mandibular position. *J Prosthet Dent*, **20**, 21–36.

8 McHorris, W. H. (1986) Centric Relation: Defined. *J Gnathology*, **5**, 5–21.

9 Nomenclature committee, Academy of denture prosthetics, Glossary of prosthodontic terms (1987). *J Prosthet Dent*, **58**, 725–762

10 Phillips, R. W., Hamilton, A. I., Jendresen, M. D. *et al.* (1986) Report of the committee on scientific investigation of the American academy of restorative dentistry. *J Prosthet Dent*, **55**, 736–772

11 Wood, G. N. (1988) Centric relation and the treatment position in rehabilitating occlusions: A physiologic approach. Part I: Developing an optimum mandibular posture. *J Prosthet Dent*, **59**, 647–651.

12 Hickey, J. C. (1964) Centric relation. A must for complete dentures. *Dent Clin North Am*, **8**, 587–600.

13 Academy of Prosthodontics (1999) Glossary of prosthodontic terms. 7th ed. *J Prosthet Dent*, **81**, 44–112

14 Kantor, M. E., Silverman, S. I., and Garfinkel, L. (1972) Centric-relation recording techniques – A comparative investigation. *J Prosthet Dent*, **28**, 593–600

15 Bansal, S., and Palaskar, J. (2009) Critical evaluation of methods to record centric jaw relation. *J Indian Prosthodont Soc*, **9**, 120–126.

16 Kapur, K. K., and Yurkstas, A. A. (1957) An evaluation of centric relation records obtained by various techniques. *J Prosthet Dent*, **7**, 770–786.

17 Myers, M. L. (1982) Centric relation records-historical review. *J Prosthet Dent*, **47**(2), 141–145.

18 Gysi, A. (1910) The problem of articulation. *Dent Cosmos*, **52**, 1–19.

19 Mohamed, A., El-Aramany, M. A., George, W. A., and Scott, R. H. (1965) Evaluation of the needle point tracing as a method for determining centric relation. *J Prosthet Dent*, **15**, 1043

20 Massad, J. J., Connelly, M. E., Rudd, K. D., and Cagna, D. R. (2004) Occlusal device for diagnostic evaluation of maxillomandibular relationships in edentulous patients: A clinical technique. *J Prosthet Dent*, **91**, 586–90.

21 Anderson, J. D., and Zarb, G. (2012) The dentures' polished surfaces, recording jaw relations, and their transfer to an articulator, in *Prosthodontic Treatment for Edentulous Patients: Complete Dentures and Implant-Supported Prostheses*, 13th edn. (eds G. Zarb, J. A. Hobkirk, S. E. Eckert, R. F. Jacob). Elsevier, St. Louis, MO, pp. 180–203.

7

Neutral-Zone Registration

Introduction

The "neutral zone" was first discribed by Dr. Wilfred Fish, who indicated that the denture's polished surface should be contoured so that it approximates the moveable muscles of the lips, cheek, and tongue [1, 2]. In addition to simply replacing missing oral tissues, complete dentures serve to structurally redefine true spaces and potential spaces within the oral cavity. Regardless of the fabrication technique used, functionally inappropriate denture tooth arrangement or physiologically unacceptable denture base volume or contour have been implicated in poor prosthesis stability and retention [1–6], compromised phonetics [7, 8], inadequate facial tissue support [8], inefficient tongue posture and function [9], and hyperactive gagging [10–13].

Directives provided for optimal facial-lingual arrangement of posterior denture teeth have varied dramatically over the profession's long history of complete denture therapy. As stated previously, the concept that posterior denture teeth should be arranged to occupy the position of their natural tooth predecessors has been put forward [7, 14–18]. Others have suggested that posterior denture teeth should be arranged directly over the crest of the edentulous ridge [19–23]. In addition, Weinberg [1], Pound [24–26], Halperin [27], Devan [28, 29], el-Gheriani [30], Lammie [31], Wright [9], Martone [32] and Campbell [33] have published subtly varying concepts

and philosophies for optimal facial-lingual arrangement of posterior denture teeth.

Recording the Physiologic Neutral Zone for Edentulous Patients

Of particular interest is use of the neutral zone [34, 35] to guide posterior denture tooth arrangement and denture base contouring. To define the neutral zone, consideration must be given to the potential denture space; that space in the edentulous mouth vacated by the natural dentition and dental supporting tissues and bounded by the tongue medially and the lips and cheeks laterally. The neutral zone resides within this potential denture space. More specifically, the neutral zone is that region where forces imposed by the tongue directed outward are neutralized by inwardly directed forces originating from the cheeks and lips during normal neuromuscular function [34]. In general, boundary conditions that define the neutral zone are developed through muscular contraction and relaxation during the various functions of mastication, phonation, deglutition, and facial expression.

To provide complete dentures that reside within the theoretically stabilizing boundary conditions of the neutral zone, careful attention must be given to the dynamic physiologic and functional nature of the edentulous oral cavity. Clinicians must understand,

Application of the Neutral Zone in Prosthodontics, First Edition. Joseph J. Massad, David R. Cagna,
Charles J. Goodacre, Russell A. Wicks and Swati A. Ahuja.
© 2017 John Wiley & Sons, Inc. Published 2017 by John Wiley & Sons, Inc.
Companion website: www.wiley.com/go/massad/neutral

identify, induce, and record the neuromuscular dynamics of the functioning oral tissues using a single static registration. Once accomplished, this information can then be applied to the 3-D construction of the definitive prosthesis.

The procedure for registering the neutral zone comprises two steps. The first step is performed during the maxillo-mandibular records appointment, which will guide the bucco-lingual positioning of the posterior prosthetic teeth; the second step is performed during the wax trial placement appointment, and aids in the development of the denture's polished surface [35]. A modeling plastic impression compound occlusal rim should be fabricated prior to the record appointment (as described Chapter 4) to accomplish the procedure in a timely manner.

Technique [35]

1) The mandibular record base with the modeling plastic impression compound occlusal rim is immersed in a warm water bath set at a temperature of 140°F (Figure 7.1a).

2) Once the modeling plastic impression compound is uniformly softened (Figure 7.1b) the mandibular record base with the occlusal rim is removed from the water bath and quickly placed in the patient's mouth (Figure 7.2). A maxillary record base is not used for this procedure because eliminating the maxillary record base eliminates the compressive forces that may arise during the recording of the neutral zone.

3) The patient is given a cup of warm water and is asked to swallow, then sip warm water and swallow again (Figure 7.3a).

(a)

(b)

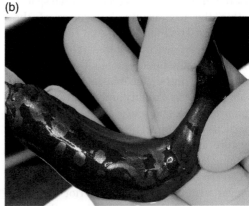

Figure 7.1 (a) Modeling plastic impression compound occlusal rim immersed in a warm water bath set at 140°F; (b) compound rim uniformly softened upon removal from the water bath.

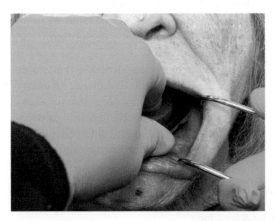

Figure 7.2 Uniformly softened modeling plastic impression compound occlusal rim inserted in the patient's mouth.

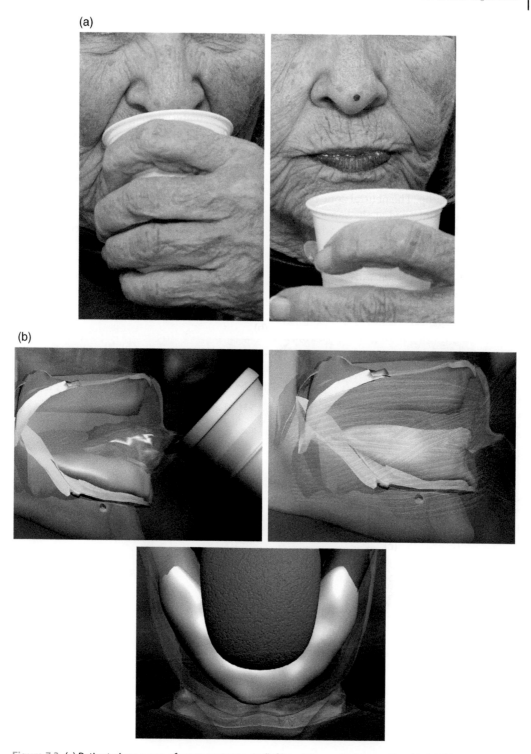

Figure 7.3 (a) Patient given a cup of warm water to sip (left) and then asked to swallow (right) (b) sipping warm water and swallowing (top left) results in muscles of the cheeks and lips functioning inward and muscles of the tongue expanding outward (top right) thereby forming the neutral zone registration (center).

4) Sipping and swallowing procedures are repeated several times. The thermoplastic rim is molded through the action of muscles of cheeks and lips moving inward and the muscles of tongue moving outward (Figure 7.3b). As the heated material cools and solidifies, the resulting volume of the modeling plastic impression compound defines the neutral zone.

5) The incisal length of the neutral zone record is compared with the patient's relaxed lower lip. The neutral zone record should be maintained at the same height as the relaxed lower lip length when the lips are parted. If the record is longer than the relaxed lower lip, a line is scribed at the level of the relaxed lower lip cooled and adjusted with a sharp blade.

6) When the modeling plastic impression compound has hardened, the neutral zone record is removed from the mouth and evaluated for accuracy (Figure 7.4). If necessary, the procedure is repeated to achieve a proper recording of the neutral zone area.

7) Excess material may be trimmed with a sharp blade (Figure 7.5). The molded compound rim can be reinserted in the mouth for verification of contours.

8) Next the neutral zone record is seated on the mandibular definitive cast and indexed lingual and facial matrices are developed around the neutral zone record using laboratory putty (Figure 7.6) [35].

This short and simple procedure uses patients' own physiological action and is

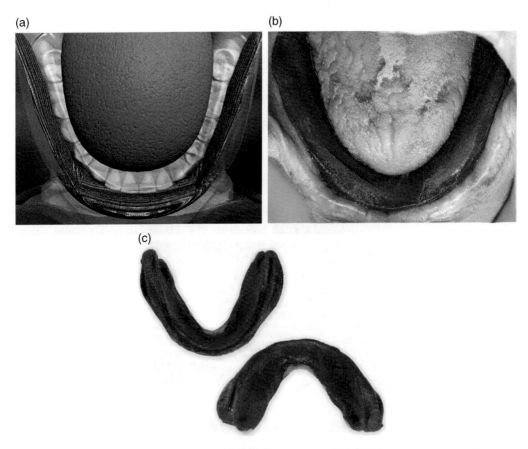

(a)

(b)

(c)

Figure 7.4 (a) Illustration of the neutral zone between tongue medially, lip labially, and cheeks buccally. (b) Clinical image demonstrating the formed registration. (c) Neutral zone registration from occlusal (left above) and lingual (right below) views.

Figure 7.5 Neutral zone registration trimmed with a sharp blade.

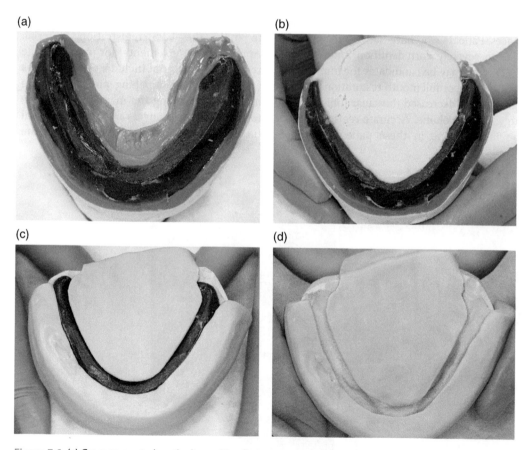

Figure 7.6 (a) Grooves created on the lingual land area serve as indices – facial and lingual vestibular spaces filled with vinyl polysiloxane impression material to facilitate complete seating of the completed indexes during tooth setup; (b) indexed lingual matrix developed around the neutral zone record using laboratory putty; (c) indexed facial matrix developed around the neutral zone record using laboratory putty; (d) neutral space visible between the facial and the lingual indexes.

therefore very repeatable. Use of this technique provides an excellent road map for optimally positioning the posterior prosthetic teeth in the mandibular denture.

Recording the Physiologic Neutral Zone for a Dentate Patient

Patients who have been edentulous for an extended period generally exhibit decreased facial muscle tone and a large tongue size, thereby affecting the retention and the stability of their prostheses. The neutral zone concept is applied to the edentulous patient to facilitate stability of the removable prosthesis and decrease food entrapment into vestibular spaces. Patients who have had significant loss of teeth, severely worn dentition, and reduced OVD (who may be candidates for immediate dentures and or full mouth restorations) often demonstrate decreased muscular tonicity and large tongue volume. Accurate registration of the neutral zone in these patients can be accomplished using a single clinical record called the cameogram.

The cameogram aids in defining appropriate bucco-lingual position of the prosthetic teeth and thickness, contours, and shape of the denture's cameo surface. During development of the cameogram, physiologic molding of impression material is accomplished so that the polished surfaces of restoration will be in physiologic harmony with the muscles of the lips, cheeks, and tongue.

Technique

High viscosity vinyl polysiloxane (VPS) impression material is injected under the lips to the full extent of the labial/buccal vestibules (Figure 7.7). The patient is instructed to perform the sequence of oral movements to mold the VPS registration material. Appropriate movements for molding the registration material include: puckering lips forward, smiling, opening and closing the mouth, and moving the mandible from side to side. These maneuvers are repeated several times. After adequate polymerization of the

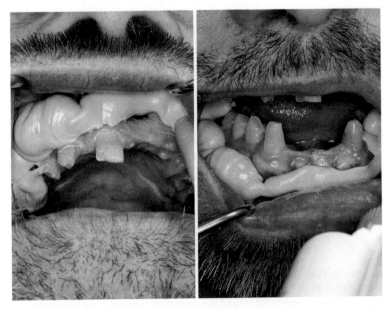

Figure 7.7 Vinyl polysiloxane (VPS) impression material injected on labial and buccal surfaces of the maxillary arch (left) and mandibular arch (right).

registration material it is carefully removed from the mouth and evaluated (Figure 7.8). Similarly, a high viscosity VPS impression material is injected on the lingual aspect of the mandibular arch to the full extent of the lingual vestibule (Figure 7.9a). Appropriate movements for molding the lingual registration include extending the tongue and moving it from left to right and licking the upper and the lower lips with the tongue. These maneuvers are repeated several times. After adequate polymerization, the registration is carefully removed from the mouth and evaluated (Figure 7.9b).

When there are significant areas of missing teeth a record base with a modeling plastic impression compound rim is fabricated to record the neutral zone in the edentulous spaces. The rim is heat softened and placed into the mouth. The patient swallows to register the inward forces of the cheeks and the outward forces of the tongue as described above ("Recording the physiologic neutral zone for edentulous patient.")

Finally, VPS putty matrices (Flexitime Easy Putty – Heraeus Kulzer) are fabricated to surround these registrations and, oriented to an indexed position on the master casts (Figure 7.10). These matrices will later be used to transfer physiologic boundaries of the registered neutral zone to cameo surface contour of transitional and/or definitive prostheses.

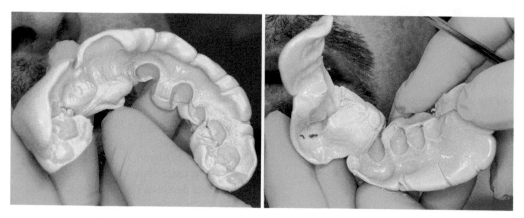

Figure 7.8 Maxillary (buccal) cameo impression (left); mandibular (buccal) cameo impression (right).

(a) (b)

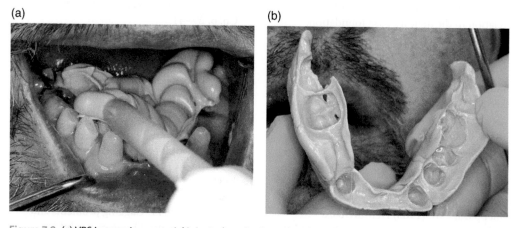

Figure 7.9 (a) VPS impression material injected on the lingual surface of the mandibular arch; (b) mandibular (lingual) cameo impression.

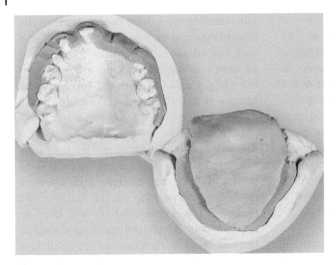

Figure 7.10 Indexed facial matrix developed around the cameo impression of the maxillary arch using laboratory putty (left). Indexed facial and lingual matrices developed around the cameo impression of the mandibular arch using laboratory putty (right).

Summary

This chapter presented the neutral zone concept and a technique for its registration. With meticulous attention to the polished surface of the prosthesis, optimal stability, retention, and comfort of the definitive complete dentures is possible.

References

1 Weinberg, L. A. (1958) Tooth position in relation to the denture base foundation. *J Prosthet Dent*, **8**, 398–405.

2 Fish, E. W. (1933) Principles of Full Denture Prosthesis. John Bale, Sons & Danielsson, Ltd, London, p. 1–8.

3 Wright, C. R. (1966) Evaluation of the factors necessary to develop stability in mandibular dentures. *J Prosthet Dent*, **16**, 414–30.

4 Brill, N., Tryde, G., Cantor, R. (1965) The dynamic nature of the lower denture space. *J Prosthet Dent*, **15**, 401–18.

5 Sheppard, I. M. (1963) Denture base dislodgement during mastication. *J Prosthet Dent*, **13**, 462–8.

6 Kuebker, W. A. (1984) Denture problems: Causes, diagnostic procedures, and clinical treatment. I. Retention problems. *Quintessence Int Dent Dig*, **15**, 1031–44.

7 Pound, E. (1954) Lost – Fine arts in the fallacy of the ridges. *J Prosthet Dent*, **4**, 6–16.

8 Fahmy, F. M., Kharat, D. U. (1990) A study of the importance of the neutral zone in complete dentures. *J Prosthet Dent*, **64**, 459–62.

9 Wright, C. R., Swartz, W. H., Godwin, W. C. (1961) Mandibular Denture Stability - A New Concept. Ann Arbor: The Overbeck Co, p. 29–41.

10 Schole, M. L. (1959) Management of the gagging patient. *J Prosthet Dent*, **9**, 578–83.

11 Morstad, A. T., Peterson, A. D. (1968) Postinsertion denture problems. *J Prosthet Dent*, **19**,126–32.

12 Means, C. R., Flenniken, I. E. (1970) Gagging – A problem in prosthetic dentistry. *J Prosthet Dent*, **23**, 614–20.

13 Kuebker, W. A. (1984) Denture problems: causes, diagnostic procedures, and clinical treatment. III/IV. Gagging problems and speech problems. *Quintessence Int Dent Dig*, **15**,1231–8.

14 Watt, D. M. (1978) Tooth positions on complete dentures. *J Dent*, **6**,147–60.

15 Watt, D. M., MacGregor, A. R. (1986) Designing Complete Dentures, 2nd ed. IOP Publishing Ltd, Bristol, p. 1–31.

16 Robinson, S. C. (1969) Physiological placement of artificial anterior teeth. *J Can Dent Assoc (Tor)*, **35**, 260–6.

17 Payne, A. G. (1971) Factors influencing the position of artificial upper anterior teeth. *J Prosthet Dent*, **26**, 26–32.

18 Murray, C. G. (1978) Re-establishing natural tooth position in the edentulous environment. *Aust Dent J*, **23**, 415–21.

19 Rahn, A. O., Heartwell, C. M. Jr. (1993) Textbook of Complete Dentures, 5th ed. Lippincott, Williams & Wilkins, Philadelphia, p. 352–6.

20 Hardy, I. R. (1942) Technique for use of non-anatomic acrylic posterior teeth. *Dent Digest*, **48**, 562–6.

21 Lang, B. R., Razzoog, M. E. (1983) A practical approach to restoring occlusion for edentulous patients. Part II: Arranging the functional and rational mold combination. *J Prosthet Dent*, **50**, 599–606.

22 Gysi, A. (1929) Practical application of research results in denture construction. *J Am Dent Assoc*, **16**, 199–223.

23 Sharry, J. J. (1974) Complete Denture Prosthodontics, 3rd ed. McGraw-Hill, New York, p. 241–65.

24 Pound, E. (1951) Esthetic dentures and their phonetic values. *J Prosthet Dent*, **1**, 98–111.

25 Pound, E., Murrell, G. A. (1973) An introduction to denture simplification. Phase II. *J Prosthet Dent*, **29**, 598–607.

26 Roraff, A. R. (1977) Arranging artificial teeth according to anatomic landmarks. *J Prosthet Dent*, **38**, 120–30.

27 Halperin, A. R., Graser, G. N., Rogoff, G. S., Plekavich, E. J. (1988) Mastering the Art of Complete Dentures. Quintessence, Chicago, p. 125.

28 Sears, V. H. (1949) Principles and technics for complete denture construction. St. Louis, Mosby, p. 103–8, 279–80.

29 DeVan, M. M., (1954) The concept of neutrocentric occlusion as related to denture stability. *J Am Dent Assoc*, **48**, 165–9.

30 el-Gheriani, A. S. (1992) A new guide for positioning of maxillary posterior denture teeth. *J Oral Rehabil*, **19**, 535–8.

31 Lammie, G. A. (1956) Aging changes and the complete lower denture. *J Prosthet Dent*, **6**, 450–64.

32 Martone, A. L. (1963) The phenomenon of function in complete denture prosthodontics. Clinical applications of concepts of functional anatomy and speech science to complete denture prosthodontics. Part VIII. The final phases of denture construction. *J Prosthet Dent*, **13**, 204–28.

33 Campbell, D. D. (1924) Full Denture Prosthesis. Mosby, St. Louis, p. 82-4.

34 Beresin, V. E., Schiesser, F. J., editors (1979). Neutral Zone in Complete and Partial Dentures, 2nd ed. Mosby, St. Louis, p. 15, 73–108, 158–83.

35 Cagna, D. R., Massad, J. J., Schiesser, F. J. (2009) The neutral zone revisited: from historical concepts to modern application. *J prosthet Dent*, **101**, 405–12.

8

Second Laboratory Procedure: Selection and Arrangement of Prosthetic Teeth

Introduction

Definitive casts should be appropriately indexed and mounted in the semiadjustable articulator using facebow and interocclusal records (Figure 8.1). Condylar inclination can be programmed using protrusive or lateral interocclusal records when a balanced occlusal scheme is desired.

Recently, the demand for esthetics has increased in the general population [1]. Hence, appropriate selection and placement of teeth is critical to successful denture therapy. Most of the information necessary for the selection of teeth should be gathered during the diagnosis and treatment-planning stage [2]. The shade selection, mold selection, and positioning and arrangement for maxillary anterior teeth may be based on the age, sex, and personality of the patient [1, 3].

Information from the esthetic blueprint wax occlusal rim and neutral zone registration are maintained by developing indexes. These indexes aid in proper arrangement of prosthetic teeth. Selection of posterior tooth form should be based on the desired occlusal scheme. Three common occlusal schemes are monoplane occlusion, lingualized occlusion, and anatomic balanced occlusion [4]. Consideration must be given to various factors such as ridge size, ridge shape, maxillo-mandibular relationship, previous denture experience, esthetic desire, general health, and the dexterity of the patient [4, 5].

The clinical skill, knowledge, judgment and experience of the dental practitioner may also play a decisive role in selecting the appropriate occlusal scheme. This chapter describes a procedure for indexing the EBP wax rim and the neutral zone record. It also discusses the selection and arrangement of prosthetic teeth.

Indexing the Esthetic Blueprint Record

The esthetic blueprint (EBP) wax occlusal rim serves as a guide for positioning the prosthetic teeth and is based on patient esthetics and phonetics. This rim conveys esthetic information: midline, high smile, line, labial inclination (lip support), tooth display in repose and smile, and orientation of the occlusal plane. The information carried by the EBP wax occlusal rim is maintained in the EBP index, which serves as a guide for developing the wax trial denture. A procedure for fabricating the EBP index is described below.

The record base and EBP wax rim are placed on the maxillary cast. Indentations are made on the base of the master cast using a rotary instrument (Figure 8.2a). An appropriate amount of laboratory putty is mixed per the manufacturer's recommendations. The putty is thoroughly kneaded, and adapted over the labial/buccal aspect of the wax rim. The putty index should capture both the land area of the

Application of the Neutral Zone in Prosthodontics, First Edition. Joseph J. Massad, David R. Cagna, Charles J. Goodacre, Russell A. Wicks and Swati A. Ahuja.
© 2017 John Wiley & Sons, Inc. Published 2017 by John Wiley & Sons, Inc.
Companion website: www.wiley.com/go/massad/neutral

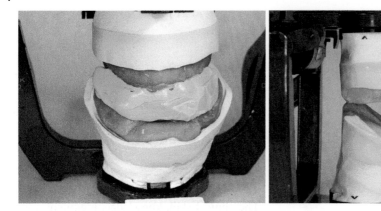

Figure 8.1 Definitive casts mounts on semiadjustable articulator using maxillo-mandibular jaw relationship records (left). Lateral view of the mounted casts (right).

(a)   (b)

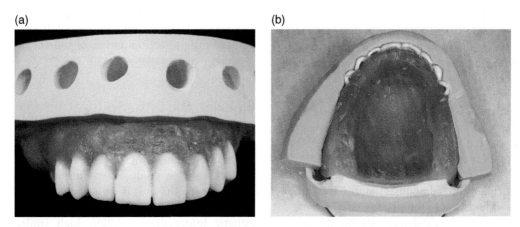

Figure 8.2 (a) Indentations made on the labial surface of the cast; (b) indexed facial matrix fits accurately on the land area of the cast.

cast and the facial contour of the wax rim so it is level with the occlusal plane of the rim. The index is removed from the cast, inspected for accuracy, and trimmed. All the markings from the wax rim should be transferred to the index. The index must fit accurately on the land area of the cast (Figure 8.2b).

Indexing the Neutral Zone Record

The neutral zone record aids in establishing the contour of the mandibular prosthesis arch form, defining the width of the occlusal surfaces, and facilitating optimal selection of the size and position of the mandibular posterior teeth [6]. The information carried by the neutral zone record should be maintained by developing an index, which subsequently serves as a guide for developing the wax trial denture. The procedure for fabricating the index is [6]:

1) The record base with the neutral zone registration is placed on the mandibular cast. Indentations are made on the base of the master cast using a rotary instrument and the facial and lingual vestibular spaces are blocked out with VPS impression material. This permits complete seating

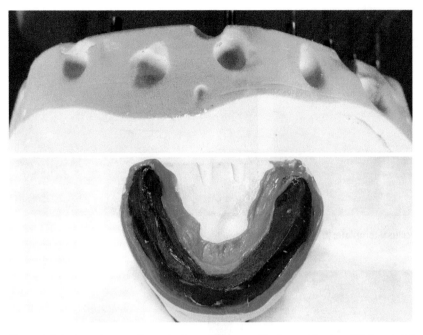

Figure 8.3 Indentations made on the base of the mandibular cast (top) and VPS material used to block out vestibular extensions (bottom).

of both indexes during teeth set up (Figure 8.3).

2) Laboratory putty is mixed per manufacturer's recommendations, thoroughly kneaded, and adapted on the lingual aspect of the neutral zone record. It is carefully molded so that it fills the tongue space completely. This putty index should capture both the land area of the cast and the lingual contour of the neutral zone record. It should be formed level with the occlusal plane of the neutral zone record.

3) A curved occlusal template may be positioned over the posterior tooth areas if a balanced occlusal scheme is desired (Figure 8.4a). A flat occlusal template can be used if a monoplane occlusal scheme is indicated (Figure 8.4b).

4) Additional putty material is mixed and adapted over the labial and buccal aspect of the neutral zone record. The putty index should capture both the land area of the cast and the facial contour of the neutral zone record. The putty should be

formed level with the occlusal plane of the neutral zone record. Following polymerization, the putty index is removed from the cast and inspected for accuracy. It should fit accurately on the land area of the cast (Figure 8.5).

Selection of Anterior Teeth

Most of the information necessary for the selection of teeth should be gathered during the diagnosis and treatment planning stage. Dental casts, photographs, old dentures, commercial guides, and / or clinically adjusted wax occlusal rims may be referenced to aid in the selection and arrangement of anterior teeth [7]. Dentists must consider the desires of the patient, esthetic demands, and functional occlusal philosophies during tooth selection in order to achieve a successful result [5].

The shade selection, mould selection, and tooth position/arrangement should harmonize with the age, sex and personality of the

(a)　　　　　　　　　　　　　　　　　(b)

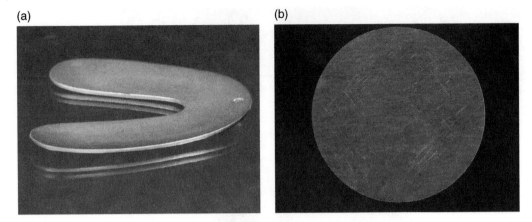

Figure 8.4　(a) Curved occlusal template; (b) flat occlusal plane.

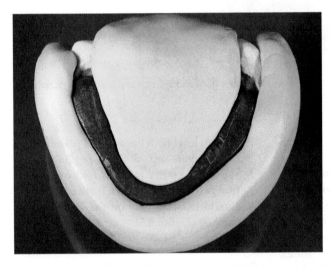

Figure 8.5　Indexed facial and lingual matrices trimmed and replaced on the cast.

patient [3]. It is essential that all pertinent information be relayed to the dental laboratory in order to achieve the desired esthetic result. In addition, prosthetic teeth may be altered by the dentist to simulate abrasion, erosion, masculinity, femininity, and so forth, during the trial denture placement appointment. The order of denture tooth arrangement is as follows:

1) Maxillary anterior teeth.
2) Mandibular anterior teeth.
3) Mandibular posterior teeth.
4) Maxillary posterior teeth.

Maxillary Anterior Teeth Arrangement

The prosthetic teeth should be set in nearly the same position that was occupied by the natural teeth, in order to achieve optimal esthetics, function, and patient comfort [3]. The EBP wax rim, EBP index, and anatomic landmarks (incisive papilla) may be used as a guide for setting the maxillary anterior teeth. In general, the incisive papilla is located between and palatal to the two maxillary central incisors [7]. The labial surface of

(a)

(b)

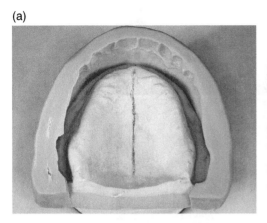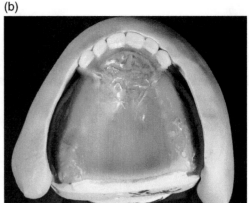

Figure 8.6 (a) The midline of the teeth should coincide with the midline marked on the EBP index and also should correlate with the position of the incisive papilla; (b) the labial aspect of the maxillary anteriors is aligned with the inner surface of the EBP index.

the central incisors is usually positioned 8–12 mm labial to the incisive papilla, depending on the skeletal relationship and amount of residual ridge resorption [8].

Prior to setting the maxillary anterior teeth, the EBP wax rim is placed on the maxillary cast. A heated instrument is used to remove wax from the rim. A central incisor is positioned carefully, using the EBP index as a guide, ensuring that the labial aspect of the central incisor approximates the inner surface of the index. The position and shape of the incisal edge is adjusted if indicated. The contralateral central incisor is set next, ensuring that it contacts the index anteriorly and that the length is adjusted appropriately. The denture tooth midline should coincide with the midline marked on the EBP index and should correlate with the position of the incisive papilla (Figure 8.6a). Next, both the lateral incisors are set. For patients desiring a natural tooth arrangement, placement of the lateral incisors can be modified (spacing, rotation, overlapping) to characterize the appearance of the denture [3]. The canines are placed in reference to the interalar line marked on the EBP index [9–11]. This line should coincide with the mesio-distal center of the canine teeth. The labial perimeter of the anterior teeth must coincide with the EBP index (Figure 8.6b).

From a frontal perspective, canines should be slightly inclined distally; with cusp tips at the same level as the central incisors, necks should be prominent, and the distal half of the labial surface should not be visible during tooth display by the patient (Figure 8.7) [1, 2].

Examples of how sex, personality, and age of the patients can be reflected in the natural and a dynamic arrangement of teeth are given in Figures 8.8a, b, and c [3]. Such characterization may be used to achieve the natural and esthetic result described in Table 8.1.

Mandibular Anterior Teeth Arrangement

Vertical and horizontal separation of the anterior teeth is recommended for complete dentures to help minimize undesirable dislodging forces and stresses on anterior ridges during mandibular movements. A degree of vertical overlap is necessary when a balanced occlusal scheme is planned (Figure 8.9) [1, 2].

The neutral zone registration is placed on the mandibular cast. The modeling plastic impression compound occlusal rim is removed from the trial record base and the lingual neutral zone index is placed on the

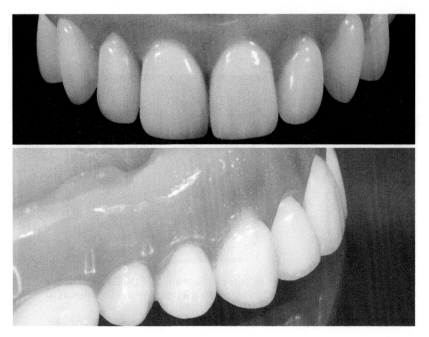

Figure 8.7 Canines are slightly distally inclined with prominent cervical thirds and their tips are at the same level occlusally as the central incisor. The distal portions of the labial surfaces are not visible from the frontal perspective when the patient displays the teeth. Frontal view (top). Profile view (bottom).

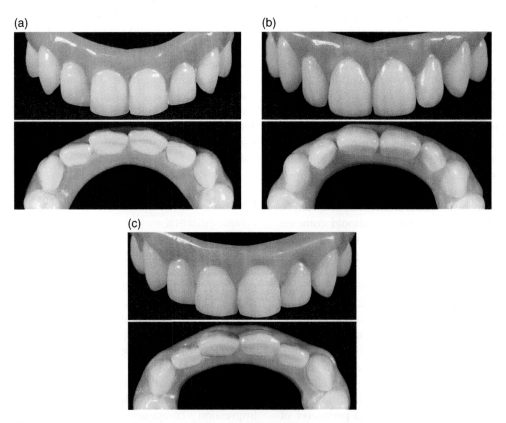

Figure 8.8 (a) Characteristics seen in male dentitions. Note sharp line angles, point angles and square shaped teeth. Frontal view (top). Occlusal view (bottom). (b) Characteristics seen in female dentitions. Note rounded line angles and point angles and oval shaped teeth. Also the lateral incisor is rotated to create a soft appearance. Frontal view (top). Occlusal view (bottom). (c) Characteristics seen in aging dentitions. Note the presence of wear facets, staining and cervical recession associated with the teeth. Frontal view (top). Occlusal view (bottom).

mandibular cast. Using the lingual neutral zone index as a guide, a wax rim is fabricated on the mandibular record base. The mandibular anterior teeth are set in wax. It is important to close the articulator periodically while setting the mandibular anterior teeth to determine appropriate vertical placement as it relates to the previously set maxillary anterior teeth. The mandibular anterior teeth should be placed within the neutral zone to achieve optimal esthetics, phonetics, and function (Figure 8.10).

Table 8.1 Amount of maxillary incisor display in repose based on the sex and age of the patient [3]

	Age and sex of the patient	Amount of maxillary central incisor display in repose (millimeters)
a	Young female	3–4 mm
b	Young male	1.5–2 mm
c	Middle-age female	1–2 mm
d	Old male	0–1 mm

Selection of Posterior Teeth

Selection of posterior tooth form is related to the choice of occlusal scheme [7]. Several factors may influence the choice of the occlusal scheme [5]:

- ridge size;
- ridge shape;
- maxillo-mandibular relationship;
- muscular coordination;
- physical or mental condition;
- previous dentures;
- complicating anatomic conditions;
- desires of the patient; and
- skill of the dentist.

Three concepts of occlusion have been used more than others for complete denture patients [4, 5]. Monoplane occlusion incorporates nonanatomic (cuspless) teeth set to a flat occlusal plane [12, 13]. Lingualized occlusion is versatile and may use anatomic or semianatomic maxillary teeth set to a flat or curved occlusal plane [14, 15]. Anatomic (balanced) occlusion

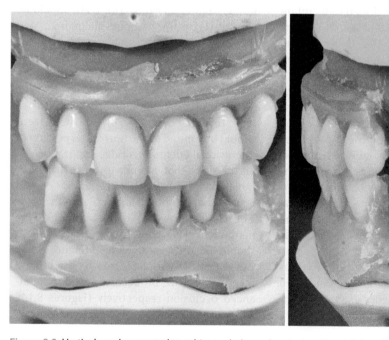

Figure 8.9 Vertical overlap created to achieve a balanced occlusion. Frontal view (left). Lateral view (right).

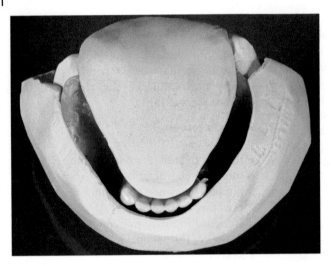

Figure 8.10 The mandibular anterior teeth placed within the neutral zone space. Note the anterior teeth are placed within the neutral space but slightly lingual to the labial index to achieve the desired horizontal overlap between the maxillary and the mandibular anterior teeth.

involves posterior teeth with greater than 30° of cuspal inclination, set to a curved occlusal plane [16–18]. Additional variations have been suggested incorporating cuspless teeth set to compensating curves or ramped second molars in order to develop balanced occlusion. Special tooth forms have also been developed to facilitate optimized lingualized occlusion. Most dentures fabricated using the neutral zone theory employ monoplane or lingualized occlusal concepts.

Traditional precepts suggest that the mandibular posterior teeth should be centered over the crest of the residual ridge. However, when accounting for the external neuromuscular forces of the edentulous oral environment (the neutral zone), mandibular posterior teeth may be most appropriately positioned buccal to the residual ridge crest. This may necessitate that the posterior teeth be set in a reverse articulation (cross bite).

Examples of this relationship versus normal horizontal articulation are illustrated for various occlusal schemes used with neutral zone theory (Figures 8.11a–d).

Mandibular Posterior Teeth Arrangement

The mandibular posterior teeth should be set within the neutral zone (Figure 8.12a). The lingual surfaces of the posterior teeth should contact the lingual extent of the neutral zone index. Mandibular posterior teeth should be placed vertically, without inclination, and their occlusal surfaces should contact the occlusal surface of the EBP wax rim. Posteriorly, the occlusal plane should be placed at the level of the middle [1] or at the junction of the middle and upper thirds of the retromolar pads [19, 20]. Prosthetic teeth should be limited to the horizontal portion of the residual edentulous ridge and not placed on the slope that ascends towards the retromolar pad (Figure 8.12b) [1]. Anterior and posterior teeth should follow a smooth curve. The central groves and the centers of the marginal ridges of all the posterior teeth should lie on a straight line (Figure 8.13). Curved or flat templates may be used to aid in development of balanced and nonbalanced occlusion respectively (Figures 8.14a and 8.14b).

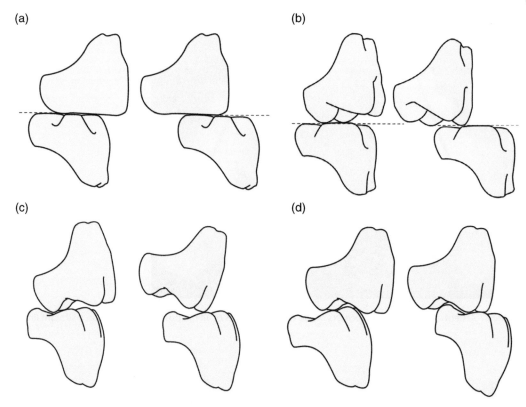

Figure 8.11 (a) Nonanatomic prosthetic teeth set in monoplane occlusion – normal (left) and reverse (right) articulation; (b) semianatomic prosthetic teeth set in monoplane occlusion – normal (left) and reverse (right) articulation; (c) Semianatomic or anatomic prosthetic teeth set to balance stamp cusps – normal (left) and reverse (right) articulation; (d) anatomic or semi anatomic prosthetic teeth set to balance stamp and supporting cusps – normal (right) and reverse (left) articulation.

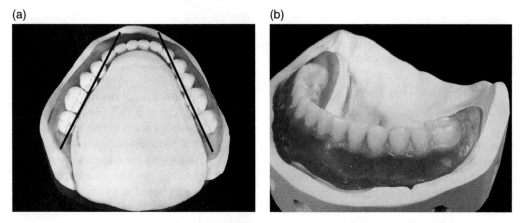

Figure 8.12 (a) The mandibular posterior teeth set within the neutral zone. Line indicates crest of the mandibular ridge; (b) The occlusal plane posteriorly placed at the middle or at the level of 2/3rd the height of the retromolar pad. Note: prosthetic teeth should not be placed on the slope of the retromolar pad.

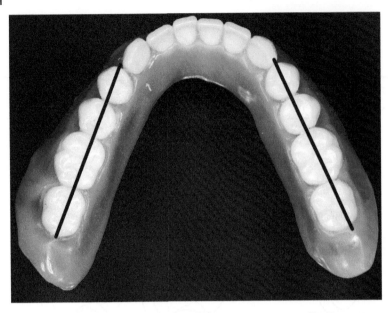

Figure 8.13 Central groves and the centers of the marginal ridges of all the posterior teeth lie on a straight line.

(a)   (b)

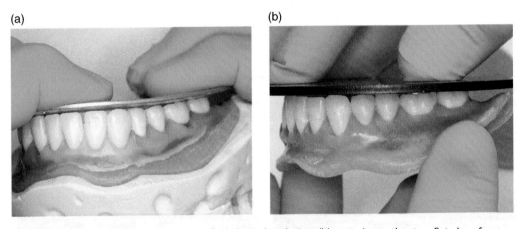

Figure 8.14 (a) posterior teeth set on curve for balanced occlusion; (b) posterior teeth set on flat plane for nonbalanced (monoplane) occlusion.

Maxillary Posterior Tooth Arrangement

This part of the procedure can be accomplished relatively quickly because previously set mandibular posterior teeth provide guidance to position maxillary posterior teeth. Wax is melted with a heated instrument and each tooth is carefully placed into position. All maxillary posterior teeth are placed within the land area and ideally with a

quarter-tooth horizontal overlap (to prevent cheek biting) (Figure 8.15). As each maxillary posterior tooth is placed in the soft wax, the articulator is immediately closed so as to achieve contact with the opposing mandibular teeth, forcing the maxillary tooth into appropriate position. Each subsequent tooth is set in this manner until all maxillary posterior teeth have been placed.

Residual ridge resorption is one of the sequelae of edentulism. The maxilla resorbs

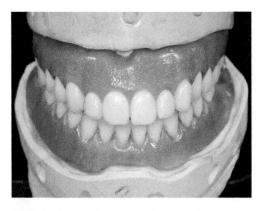

Figure 8.15 Posterior teeth set in normal articulation with appropriate horizontal overlap.

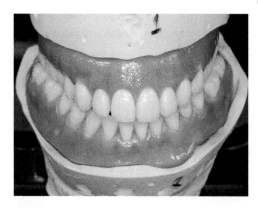

Figure 8.16 Maxillary posterior teeth set in cross bite (reverse articulation).

inward and backward, and the mandible resorbs downward and outward, creating conditions conducive to reverse articulation [21–25]. As previously stated, the neutral zone record may indicate that the mandibular teeth be set further buccal than the ridge crest. In these patients, setting the maxillary posterior teeth in normal horizontal articulation may compromise patient comfort, function, and denture stability [26]. In these situations it becomes necessary to set the maxillary posterior teeth in reverse articulation to achieve the desired stability of the complete denture (Figure 8.16) [27].

Upon completion, the position of all maxillary and mandibular posterior teeth and the occlusion are evaluated and verified. The articulator may be moved in excursions to reveal any abnormalities that may exist.

Tooth Selection and Arrangement for the Partially Edentulous Patient

Teeth selection is performed for partially edentulous patients in a manner similar to that described for edentulous patients. For a dentate patient (indicated for complete dentures) the teeth from the cast are removed and its surface topography is modified to represent the expected resorptive changes in contour following tooth extraction (rule of thirds) [28]. Trial denture placement may only be accomplished for the edentulous areas, when indicated. Tooth arrangement is performed as follows.

VPS putty matrices (Flexitime Easy Putty – Heraeus Kulzer), fabricated to surround the neutral zone records (described in Chapter 7), are used to transfer the physiologic boundaries of the registered neutral zone and aid in arrangement of teeth and development of the cameo surfaces for transitional and / or definitive prostheses (Figures 8.17a and b). The labial perimeter established by the anterior teeth should correspond to the VPS matrix (Figure 8.17c).

The maxillary posterior teeth are arranged based on the VPS matrix and by referencing the position of the mandibular posterior teeth (Figure 8.17d).

When the patient does not desire a mandibular restoration, the maxillary posterior teeth should be set in contact with the remaining mandibular teeth (Figure 8.17e). Once the occlusal plane of the maxillary teeth is established, restoration of the missing mandibular teeth can be considered.

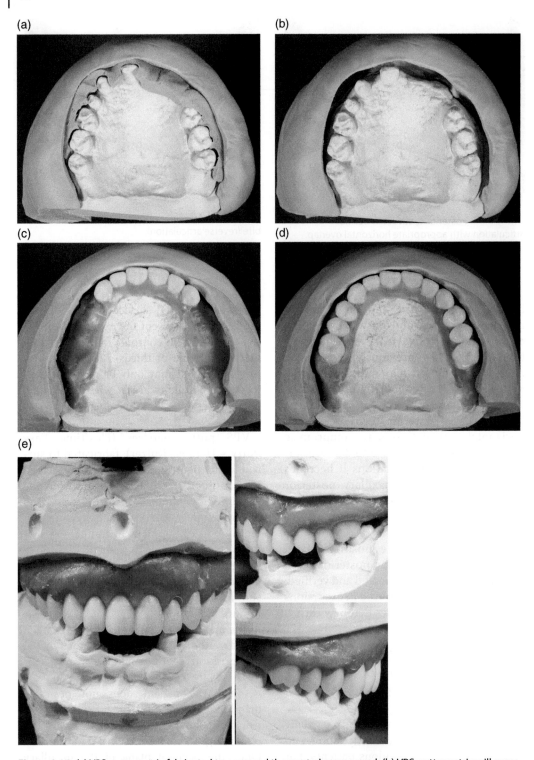

Figure 8.17 (a) VPS putty matrix fabricated to surround the neutral zone record; (b) VPS putty matrix will serve as a reference during teeth arrangement; (c) the labial perimeter established by the combined anterior teeth is consistent with the VPS matrix; (d) maxillary posterior teeth arrangement determined by the VPS matrix and the arrangement of the mandibular posterior teeth; (e) maxillary teeth arranged to articulate with the opposing mandibular teeth. Frontal view (left). Left lateral view (top, right). Right lateral view (bottom, right).

All the maxillary posterior teeth are placed within the land area of the cast and ideally with a quarter-tooth horizontal overlap (to prevent cheek biting). As each maxillary posterior tooth is placed in the soft wax, the articulator is closed into occlusion with the opposing mandibular teeth moving the maxillary tooth into proper occlusion.

Summary

Use of the EBP wax rim and the neutral zone rim for setting the prosthetic teeth has been described. Once the teeth are set and their position and arrangement are verified, the wax trial dentures are contoured and festooned in preparation for the trail denture placement appointment.

References

1 Lefebvre, C., and Ivanhoe, J. R. (2009) Tooth arrangement, in *Textbook of Complete Dentures*, 6th edn. (eds. A. O. Rahn, J. R. Ivanhoe, K. D. Plummer). People's Medical Publishing House, Shelton, CT, pp. 198–215.

2 Fenton, A. H., Chang, T.-L. (2012) The occlusal surfaces: The selection and arrangement of prosthetic teeth, in *Prosthodontic Treatment for Edentulous Patients: Complete Dentures and Implant-Supported Prostheses*, 13th edn. (eds. G. Zarb, J. A. Hobkirk, S. E. Eckert, and R. F. Jacob). Elsevier, St. Louis, MO, pp. 204–229.

3 Frush, J. P., and Fisher, R. D. (1958) The dynesthetic interpretation of the dentogenic concept. *J Prosthet Dent*, 8, 558–581.

4 Tarazi, E., and Ticotsky-Zadok, N. (2007) Occlusal schemes of complete dentures – a review of the literature. *Refuat Hapeh Vehashinayim (1993)*, 24, 56–64, 85–86.

5 Parr, G. R., and Loft, G. H. (1982) The occlusal spectrum and complete dentures. *Compend Contin Educ Dent*, 3, 241–250.

6 Cagna, D. R., Massad, J. J., and Schiesser, F. J. (2009) The neutral zone revisited: from historical concepts to modern application. *J Prosthet Dent*, 101, 405–412.

7 Kapur, K. K. (1973) Occlusal patterns and tooth arrangements, in *International Prosthodontic Workshop on Complete Denture Occlusion* (eds. B. R. Lang and C. C. Kelsey), University of Michigan School of Dentistry, Ann Arbor, MI, pp. 145–172.

8 Ortman, H. R. (1979) Relationship of the incisive papilla to the maxillary central incisors. *J Prosthet Dent*, 42, 492–496.

9 Hoffman, W., Bomberg, T. H., and Hatch, R. A. (1986) Interalar width as a guide in denture tooth selection. *J Prosthet Dent*, 55, 219–221.

10 Mavroskoufis, F., and Ritchie, G. M. (1981) Nasal width and incisive papilla as guides for the selection and arrangement of maxillary anterior teeth. *J Prosthet Dent*, 45, 592–597.

11 Wehner, P. J., and Hickey, J. C., Boucher, C. O. (1967) Selection of artificial teeth. *J Prosthet Dent*, 18, 222–232.

12 DeVan, M. M. (1954) The concept of neutrocentric occlusion as related to denture stability. *J Am Dent Assoc*, 48, 165–169.

13 Brudvik, J. S., and Wormley, J. H. (1968) A method of developing monoplane occlusions. *J Prosthet Dent*, 19, 573–580.

14 Parr, G. R., and Ivanhoe, J. R. (1996) Lingualized occlusion. An occlusion for all reasons. *Dent Clin North Am*, 40(1), 102–112.

15 Clough, H. E., Knodle, J. M., Leeper, S. H., *et al.* (1983) A comparison of lingualized occlusion and monoplane occlusion in complete dentures. *J Prosthet Dent*, 50(2), 176–179.

16 Landa, J. S. (1962) Biologic significance of balanced occlusion and balanced articulation in complete denture service. *J Am Dent Assoc*, 65, 489–494.

17 Sutton, A. F., and McCord, J. F. (2007) A randomized clinical trial comparing anatomic, lingualized, and zero-degree posterior occlusal forms for complete dentures. *J Prosthet Dent*, 97(5), 292–298.

18 Meyer, F. S. (1934) A new, simple and accurate technic for obtaining balanced and functional occlusion. *J Am Dent Assoc*, **21**, 195–203.

19 Boucher, C. O. (1964) *Swenson's Complete Dentures*, 5th edn. Mosby, St. Louis, MO, pp. 246–251.

20 Hall, W. A. (1958) Important factors in adequate denture occlusion. *J Prosthet Dent*, **8**, 764–775. doi: 10.1016/0022-3913(58)90096-9

21 Gysi, A. (1930) Occlusion and the cross-bite set-up, in *Prosthetic Dentistry – An Encyclopedia of Full and Partial Denture Prosthesis* (ed. I. G. Nichols). Mosby, St. Louis, MO, pp. 337–342.

22 Pound, E. (1954) Lost – fine arts in the fallacy of the ridges. *J Prosthet Dent*, **4**, 6–16.

23 Boucher, C. O. (1964) *Swenson's Complete Dentures*, 5th edn. Mosby, St. Louis, MO, pp. 215–286.

24 Sicher, H. (1965) *Oral Anatomy*, 4th edn. Mosby, St. Louis, MO, p. 201.

25 Davis, D. M. (1997) Developing an analogue / substitute for mandibular denture-bearing area, in *Boucher's Prosthodontic Treatment for Edentulous Patients*, 11th edn. (eds G. A. Zarb, C. L. Bolender, and G. E. Carlsson). St. Louis: Mosby, pp. 162–181.

26 LaVere, A. M., and Freda, A. L. (1972) Artificial tooth arrangement for prognathic patients. *J Prosthet Dent* **28**(6), 650–654.

27 Massad, J. J., William, J. D., June, R., *et al.* (n.d.) Rationale for adhesive in complete denture therapy. *Dent Today*, http://www.dentalcare.es/media/es-ES/research_db/pdf/fixo/massad.pdf (accessed February 12, 2017).

28 Jerbi, F. C. (1996) Trimming the cast in the construction of immediate dentures. *J Prosthet Dent*, **16**, 1047–1053.

9

Trial Placement Appointment

Trial Placement

The trial placement appointment is a very significant appointment in denture fabrication [1–3]. It is the dental practitioner's last chance to ensure that the definitive prostheses will satisfy the esthetic, phonetic, and functional needs of the patient. During the trial placement appointment, the position, shade, and mold of the prosthetic teeth, lip, and facial support, occlusal vertical dimension, centric occlusion and the horizontal and vertical relationship of the casts are evaluated on the articulator [1–3]. The patient and the dental practitioner have a final opportunity to view the esthetic design of the definitive prostheses prior to denture processing. If there is a concern or issue, it must be corrected during this appointment. The patient and the dental professional should be completely satisfied before proceeding with the irreversible procedure of processing dentures.

It is important to invite both the patient and the patient's significant other (who may play a very important role in the patient's decisions) to review the result of the previous denture appointments. This involvement of the significant other may help the patient adapt to the new dentures post denture delivery. The time taken for this appointment can vary from minutes to hours, depending on the thoroughness with which the previous procedures were accomplished and the personality type of the patient. Always remember that it is much easier and less expensive to make adjustments

while the teeth are set in wax, and it is impossible or too expensive to make the adjustments once the denture is processed.

It is critical that the articulator, casts, and the wax up, presented in front of the patient be clean, neat, and tidy. Before proceeding with the trial placement appointment, the record bases should be evaluated (Figure 9.1) and adjusted to ensure proper fit and comfort. If the record bases are not retentive, an appropriate adhesive may be used to keep them stable during the appointment. It should be explained to the patient that the trial prosthesis is not their actual dentures and that the trial dentures may not fit as well as the definitive prosthesis because they are fabricated with block outs to prevent any damage to the definitive cast. Also, advise patients that the color of acrylic resin in the definitive prostheses will be more natural and esthetic than the wax trial denture. The patient can be shown shade tabs to help determine the final acrylic resin shade and to defray concerns regarding the color of the baseplate wax.

Evaluation of Esthetics

Patients should be standing in a normal postural position and should be 6–8 feet away while evaluating their smile and esthetics with the trial prostheses. Evaluating the patient while they are sitting in the dental chair does not give a complete picture of their smile. Many practitioners ask the patient to

Application of the Neutral Zone in Prosthodontics, First Edition. Joseph J. Massad, David R. Cagna, Charles J. Goodacre, Russell A. Wicks and Swati A. Ahuja.
© 2017 John Wiley & Sons, Inc. Published 2017 by John Wiley & Sons, Inc.
Companion website: www.wiley.com/go/massad/neutral

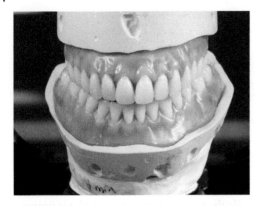

Figure 9.1 Evaluation of the teeth arrangement on the articulator.

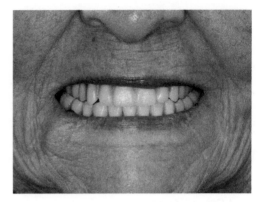

Figure 9.2 Position, shade and mold of prosthetic teeth evaluated during the trial placement appointment.

walk away and then turn around and smile to get a natural perception of their smile whether it is pleasing or displeasing and then modifications are made as necessary. The midline, occlusal plane, individual tooth positions, facial and lip support and shade, size and shape of prosthetic teeth should be assessed during this appointment (Figure 9.2 and Figure 9.3a–c). The dentist can also include individual characterizations (such as diastemas, overlaps, wear facets, and stains) to improve the esthetics of the prostheses further.

Most patients give a big broad smile on instruction but some are unable to smile themselves. These can be assisted by asking them to say an exaggerated "E." Observation of the esthetic dental display as the patient progresses from repose to full animation provides valuable information (Figure 9.3d). As previously stated, it is important to allow patients to evaluate, view, and discuss the trial dentures with their significant others. They should be given a hand mirror, so that they can view and express their likes and dislikes regarding the trial prostheses (Figure 9.3e). Patients who are shy can be asked specific questions regarding their smile, shade, and mold of prosthetic teeth, teeth display, and teeth arrangement. All patient concerns should be noted and it should be explained that some factors, such as wrinkles and deep soft tissue folds, cannot be addressed with removable dental prostheses

alone. It is important to explain to patients that all the necessary steps to enhance their appearance have been taken. Most patients understand and adjust to complete dentures if they are educated early but it is a lost battle if we bring forth such information late, at the time of denture placement.

Evaluation of Phonetics

Phonetic assessment helps to evaluate the position of the prosthetic teeth and the space between them while the patient is speaking. Phonetics is also an established method for determining and evaluating the occlusal vertical dimension [4–8]. In this process, the patient is instructed to perform specific phonetic maneuvers and the dentolabial and interincisal approximations are observed. Labiodental / fricative / "F" and "V" sounds help assess the relationship of the maxillary anterior teeth to the lower lip. Ideally the incisal edges of the maxillary anterior teeth should lightly touch the wet-dry line of the lower lip while making the "F" and "V" sounds (Figure 9.4a). These sounds help determine if the maxillary anterior teeth are too long, too short, or too labially or lingually positioned [9].

Linguo-alveolar / sibilant / "S" sounds ("sixty-six," "Mississippi") are made with the valve formed by the tip of the tongue and anterior part of hard palate [9]. Sibilant sounds help

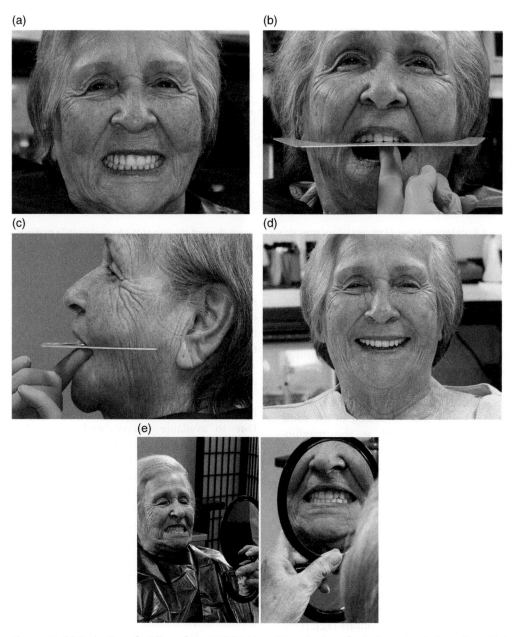

Figure 9.3 (a) Evaluation of midline of wax trial dentures; (b) evaluation of anterior occlusal plane of wax trial dentures; (c) evaluation of posterior occlusal plane of wax trial dentures; (d) evaluation of patient's smile with wax trial dentures; (e) Patient evaluating their smile with wax trial dentures.

evaluate the height of the maxillary and mandibular anterior teeth, the interocclusal distance, and the labio-lingual positioning of the maxillary anterior teeth [9]. The upper and lower anterior prosthetic teeth should approach end to end but not touch and should be visible while making these sounds (Figure 9.4b). If these sounds appear muffled, then the prosthetic teeth should be repositioned in wax to enable clear speech [9].

(a)

(b)

Figure 9.4 (a) Incisal edges of the maxillary anterior teeth touch the wet-dry line of the lower lip while making the "F" and "V" sounds; (b) the upper and lower anterior prosthetic teeth approach end to end but not touch while making the "S" sounds.

Evaluation of Occlusal Vertical Dimension (OVD)

If the patient has an existing set of dentures, a Boley gauge or a caliper can be utilized to compare the OVD of the new wax trial dentures with the existing dentures (Figure 9.5a). The OVD should be evaluated in the same way it was determined at the maxillomandibular jaw relationship records appointment [10]. It can be verified again using esthetic and phonetic assessments [4–8]. Esthetic assessment of OVD includes ensuring all the three thirds of the face (upper, middle, and lower third) are nearly equal (Figure 9.5b) [11]. The patient should appear relaxed, comfortable, and not overclosed or stretched open with the trial dentures placed in the mouth. There should be an optimal amount of interocclusal separation when the patient makes the sibilant sounds [12]. If the opposing teeth contact during sibilant sounds, the OVD may be excessive. The wax trial denture should be transferred to the articulator. The incisal guide pin is then raised to correspond with the planned reduction in OVD. Wax of the mandibular trial denture is warmed and teeth are intruded into the wax by closing the articulator to incisal pin contact. If the required reduction of OVD is greater than 3–4 mm, a new CR

record is indicated. The prosthetic teeth of the mandibular wax trial denture should be removed, as needed, to provide space for the interocclusal registration procedure. The mandibular cast is remounted in the articulator.

If an increase in OVD greater than 3–4 mm is required, a new CR record is indicated. Due to available interocclusal space, the prosthetic teeth do not require removal prior to making the CR record. The mandibular cast is remounted in the articulator.

When correcting the OVD of the wax trial dentures, any change in antero-posterior relation of the teeth and tilting of the teeth should also be corrected. After correction, the OVD the wax trial dentures must be placed in the mouth and reevaluated.

Evaluation of Centric Contact Position

It is very important to assess the accuracy of the CR record [2, 3]. The CR record can be verified by intraoral examination of denture occlusion and comparing it with the occlusion established on the articulator (Figure 9.6). Record bases should be retentive and stable to evaluate occlusion. The mandibular record base usually needs

(a)

(b)

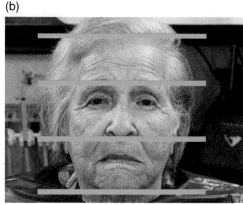

Figure 9.5 (a) Verification of OVD with wax trial dentures in the mouth; (b) the three-thirds of the face (upper third, middle third, and lower third) are nearly equal.

to be stabilized manually by the dental practitioner to prevent movement of the trial base away from the denture-bearing tissues and ensure both the base and the mandible move together as one unit. If the maxillary base is unstable it should also be stabilized manually. Denture adhesive may be used to stabilize the trial bases when indicated. The dentist can guide patient into the CR position (chin-point guidance [13] / bimanual manipulation [14]) or train patients to passively close in the CR position by asking them to place the tongue on the roof of the mouth as far posterior as possible [15]. The first contact of the teeth is observed. All the teeth should contact simultaneously with no occlusal interferences in the CR position.

Due to the use of a central bearing point devise during the interocclusal records appointment, the need to reassess cast mounting accuracy at this point is generally unnecessary.

External Impressions

Fish stated that every surface (including the polished surface) of the complete denture should fit some part of the oral tissues or some part of the opposing dentition [16, 17]. The polished / cameo surface / external surface is the surface that contacts the tongue,

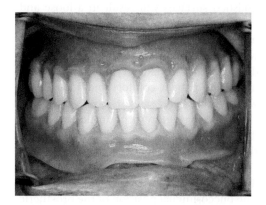

Figure 9.6 Verification of CR with wax trial dentures.

lips and the cheek [17]. The polished surface should not be developed arbitrarily by the technician. It should be appropriately contoured and defined by muscle action to conform to the shape and function of the tongue, lips, and cheeks [17, 18]. The musculature of lips, cheek, and tongue exerts an elastic pressure on the polished surfaces (when appropriately contoured) retaining the denture in place rather than dislodging it (Figure 9.7) [18]. Optimally contoured polished surfaces permit the musculature to stabilize the bolus, and, at the same time, stabilize the dentures [18]. Even in the passive state, the weight of the musculature contacting the optimally developed polished surfaces will aid in the retention of the complete denture [19]. Inappropriately contoured

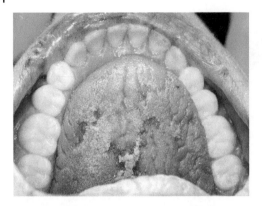

Figure 9.7 The musculature of lips, cheek, and tongue exerts an elastic pressure on the polished surfaces of the denture.

polished surfaces will result in a loose and unstable denture [18, 19]. Proper development of the polished surface not only helps with the retention and stability of complete dentures but also aids in improving speech, increasing patient comfort and adaptability with dentures.

Advanced ridge resorption decreases the denture foundation area and its influence on the stability and retention of the prostheses [18]. When the denture foundation area decreases, the polished surface area increases. The bucco-lingual position of the prosthetic teeth and the shape and form of the polished surface play a more crucial role in maintaining the stability of the complete denture [18]. The neutral zone is an area of equilibrium in the mouth where the outward forces applied by the tongue are neutralized by the inward forces applied by the lips and the cheeks [20]. These forces are developed through muscular contraction during chewing, speaking, and swallowing [20]. These forces vary in magnitude and direction within patients from time to time and also from patient to patient [19, 20]. Accurate determination of the location of the neutral zone requires two steps – a special clinical record to determine the bucco-lingual position of the prosthetic teeth on the wax trial denture (described in Chapter 6) and an additional impression procedure (external impression) for accurate registration of the

cameo surface of the denture [18]. The positioning of prosthetic teeth in the wax trial denture is accomplished using neutral zone record generated matrices (described in Chapter 8). Once the teeth are set in the wax trial denture, the patient is scheduled for an external impression procedure.

This impression, called an external impression / cameo impression / neutral zone impression, involves placing registration material on the facial, lingual, palatal aspects of the wax trial denture, including the area between the cervical aspects of the prosthetic teeth and peripheral borders of the trial dentures [19]. Several variations of the technique for making an external impression have been reported in the literature as an aid in denture construction or for use as a diagnostic tool while performing denture adjustment. The external impression records the action of the lips, the cheeks, and tongue and helps define the thickness, contours, and shape of the cameo surface [19]. During making of an external impression, physiologic molding of the impression material is accomplished such that the polished surfaces are compatible with muscle action [19].

Technique

External impressions should be made for one arch at a time. The wax trial denture is placed in the mouth and the patient is trained to perform the movements necessary to mold the registration material appropriately. In preparation for the external impression procedure, the baseplate wax apical to the denture teeth is removed carefully all the way down to the border of the wax trial denture. Baseplate wax is removed from the buccal, labial, and lingual / palatal flanges of both the maxillary and the mandibular wax trial denture (Figure 9.8a–b).

Zinc oxide eugenol paste, waxes, elastomeric impression materials or soft denture relining materials can be used as a registration material for this procedure. Vinyl polysiloxane (VPS) impression materials are simple, quick to use, and are available in various viscosities. A suitable VPS adhesive is

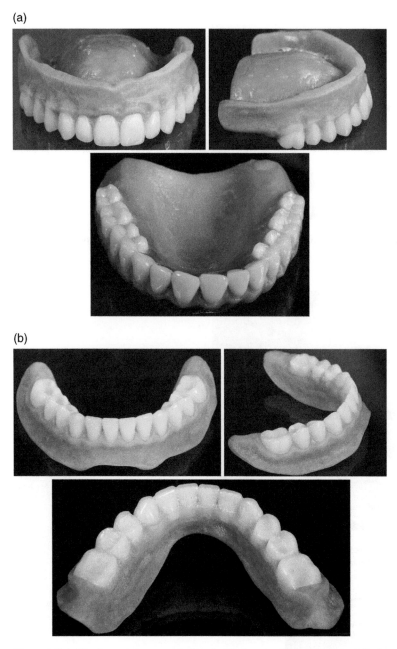

Figure 9.8 (a) The baseplate wax apical to the denture teeth removed carefully from buccal, labial and palatal flanges of the maxillary wax trial dentures – frontal view (left), occlusal view (center), right lateral view (right); (b) the baseplate wax apical to the denture teeth removed carefully from buccal, labial and lingual flanges of the mandibular wax trial dentures – frontal view (left), lingual view (center), right lateral view (right).

applied to the trial denture in the regions of wax removal to aid in the retention of the registration material (Figure 9.9). Generally, the buccal and the labial polished surfaces are registered first followed by the lingual / palatal surfaces. Light viscosity VPS impression material is applied to the trial denture in the regions of wax removal (Figure 9.10). The

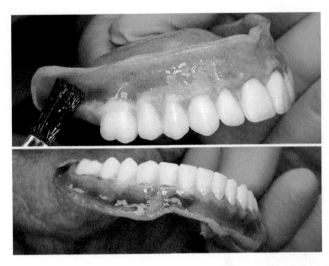

Figure 9.9 Adhesive applied to the polished surface of the wax trial dentures – maxillary denture (top), mandibular dentures (bottom).

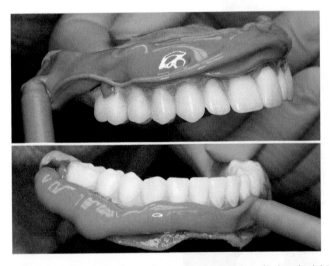

Figure 9.10 Vinyl polysiloxane impression material applied to the labial and buccal surfaces of the trial dentures in the regions of wax removal – maxillary denture (top), mandibular denture (bottom).

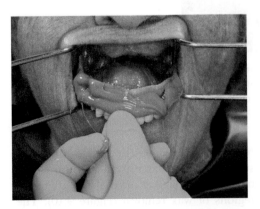

Figure 9.11 Lips and cheeks are retracted to prevent displacement of material during insertion.

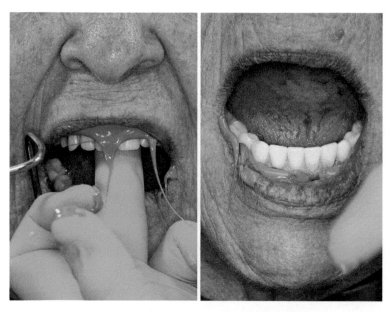

Figure 9.12 Trial denture with the VPS material inserted in the mouth – maxillary denture (left), mandibular denture (right).

lips and cheeks are retracted and the wax trial denture is placed cautiously in the patient's mouth, with particular care being taken to avoid displacing the registration material (Figures 9.11 and 9.12). The patient is instructed to perform the sequence of oral movements to mold the registration material physiologically.

Appropriate movements for physiologically molding the labial and the buccal polished surfaces of both the maxillary and the mandibular wax trial denture include: puckering lips forward, smiling, opening, and closing the mouth, and moving the mandible from side to side (Figures 9.13a–c and 9.14). After adequate polymerization of the registration material, the wax trial denture is carefully removed from the mouth and evaluated (Figures 9.15 and Figure 9.16). The lingual / palatal surfaces are registered in the same manner as discussed above.

Appropriate movements for physiologically molding the palatal polished surface of the maxillary wax trial denture include instructing the patient to sip and swallow water, and perform sibilant and fricative phonetics (Figure 9.17). Appropriate movements for physiologically molding the lingual polished surface of the mandibular wax trial denture include instructing the patient to sip and swallow water, extend their tongue and move it from left to right and lick the upper and the lower lips with the tongue (Figure 9.18) [19]. These maneuvers are repeated several times. After adequate polymerization of the registration material the wax trial denture is carefully removed from the mouth and evaluated (Figure 9.19 and Figure 9.20). Excess impression material is removed carefully with a sharp blade and minimal wax is added to blend seal the material to the cameo surfaces. The wax dentures are then related to the casts and readied for processing (Figure 9.21). It is interesting to compare general cameo surface contours of the dentures typically taught in dental schools and laboratory training facilities and the denture that result from the physiologically molded external impression procedure.

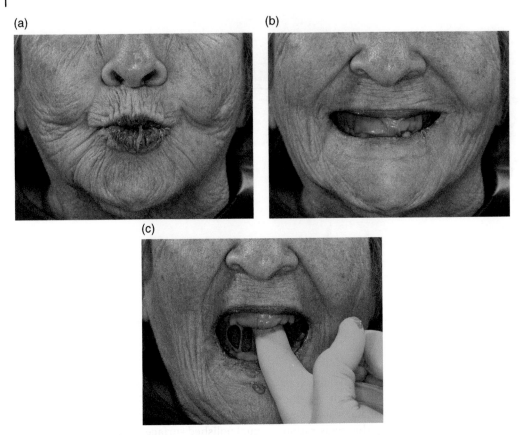

Figure 9.13 Pt movements for physiologically molding the labial and the buccal polished surfaces of the maxillary wax trial denture. (a) Pucker lips forward. (b) Full smile. (c) Open mouth wide.

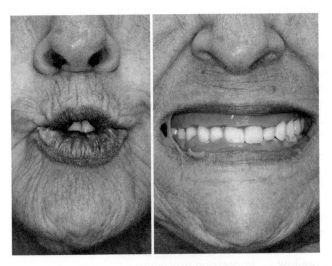

Figure 9.14 Patient movements for physiologically molding the labial / buccal surface of the mandibular wax trial denture – puckering of lips (left), smiling (right).

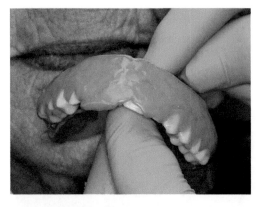

Figure 9.15 External impression of the labial and buccal surfaces of maxillary wax trial denture.

Figure 9.16 External impression of the labial and buccal surfaces of the mandibular wax trial denture.

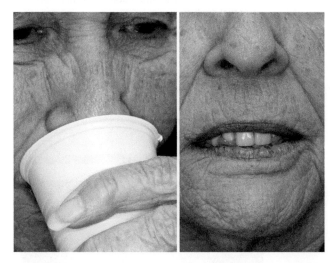

Figure 9.17 Pt movements for physiologically molding the palatal polished surface of the maxillary wax trial denture. Patient instructed to drink a sip of water and then swallow (left). Patient makes fricative sounds (right).

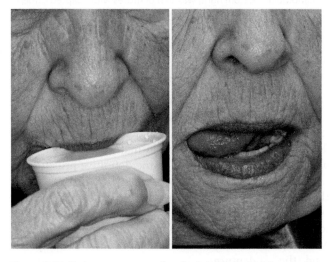

Figure 9.18 Patient movements for physiologically molding the lingual polished surface of the mandibular wax trial denture. Patient instructed to drink a sip of water and then swallow (left). Patient extends the tongue and moves it from left to right (right).

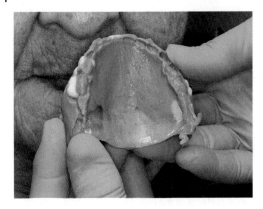

Figure 9.19 External impression of the palatal surface of the maxillary wax trial denture.

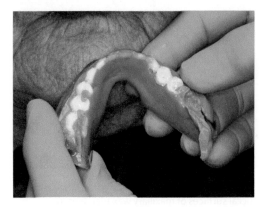

Figure 9.20 External impression of the mandibular wax trial denture.

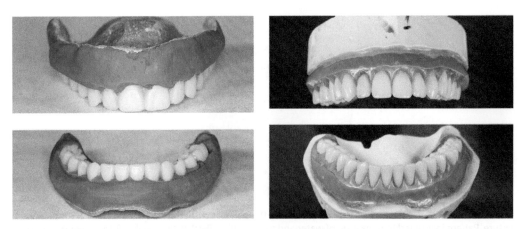

Figure 9.21 Maxillary (top left) and mandibular (bottom left) external impressions trimmed appropriately with a sharp blade. Maxillary (top right) and mandibular (bottom right) wax trial dentures with adapted cameo surfaces ready to be processed.

Summary

The trial placement appointment is very significant because it is the dental practitioner's last chance to ensure that the definitive prostheses will satisfy the esthetic, phonetic, and functional needs of the patient. During the trial placement appointment, the position of denture teeth, lip, and facial support, occlusal vertical dimension, centric occlusion, the horizontal and vertical relationship of the casts on the articulator, and the postpalatal seal are evaluated before making external impressions.

External impression of the polished surface of the wax trial denture is made with a suitable registration material and this form is carried through the definitive prostheses. It is critical to form the polished surfaces as meticulously as the impression and the occlusal surfaces. The polished surfaces play a very important role in maintaining the stability and retention of the complete denture.

References

1 Travaglini, E. A. (1980) Verification appointment in complete denture therapy. *J Prosthet Dent*, **44**, 478–483.

2 Plummer, K. D. (2009) Trial insertion appointment, in *Textbook of Complete Dentures*, 6th edn. (eds. A. O. Rahn, J. R. Ivanhoe, and K. D. Plummer). People's Medical Publishing House, Shelton, CT, pp. 217–226.

3 Fenton, A. H., and Chang, T.-L. (2012) *The try-in appointment, in Prosthodontic Treatment for Edentulous Patients*: *Complete Dentures and Implant-Supported Prostheses*, 13th edn. (eds. G. Zarb, J. A. Hobkirk, S. E. Eckert, and R. F. Jacob). Elsevier, St. Louis, MO, pp. 230–254.

4 Toolson, L. B., and Smith, D. E. (1982) Clinical measurement and evaluation of vertical dimension. *J Prosthet Dent*, **47**, 236–241.

5 Pound, E. (1978) The vertical dimension of speech: the pilot of occlusion. *J Calif Dent Assoc*, **6**, 42–47.

6 Hellsing, G., and Ekstrand, K. (1987) Ability of edentulous human beings to adapt to changes in vertical dimension. *J Oral Rehabil*, **14**, 379–383.

7 Rivera-Morales, W. C., and Goldman, B. M. (1997) Are speech-based techniques for determination of occlusal vertical dimension reliable? *Compend Contin Educ Dent*, **18**, 1214–1215, 1219–1223.

8 Silverman, M. M. (2001) The speaking method in measuring vertical dimension. *J Prosthet Dent*, **85**, 427–431. Originally published in 1952.

9 Rothman, R. (1961) Phonetic considerations in denture prosthesis. *J Prosthet Dent*, **11**, 214–223.

10 Niswonger, M. E. (1934) The rest position of the mandible and the centric relation. *J Am Dent Assoc*, **21**, 1572–1582.

11 Misch, C. E. (2015) *Dental Implant Prosthetics. A Maxillary Denture with Modified Occlusal Concepts Opposing an Implant Prosthesis*, 2nd edn. Elsevier, St. Louis, MO, pp 938–965.

12 Burnett, C. A., and Clifford, T. J. (1993) Closest speaking space during the production of sibilant sounds and its value in establishing the vertical dimension of occlusion. *J Dent Res*, **72**, 964.

13 Shafagh, I. and Amirloo, R. (1979) Replicability of chinpoint-guidance and anterior programmer for recording centric relation. *J Prosthet Dent*, **42**, 402–404.

14 Dawson, P. E. (1979) Centric relation – its effect on occluso-muscle harmony. *Dent Clin North Am* **23**, 169–180.

15 Bissasu, M. (1999) Use of the tongue for recording centric relation for edentulous patients. *J Prosthet Dent*, **82**, 369–370.

16 Fish, E. W. (1933) Using the muscles to stabilize the full lower denture. *J Am Dent Assoc*, **20**, 2163–2169.

17 Fish, E. W. (1964) *Principles of Full Denture Prosthesis*, 6th edn. Staples Press, London, pp. 36–37.

18 Beresin, V. E., and Schiesser, F. J. (1976) The neutral zone in complete dentures. *J Prosthet Dent*, **36**, 356–367.

19 Cagna, D. R., Massad, J. J., Schiesser, F. J. (2009) The neutral zone revisited: from historical concepts to modern application. *J Prosthet Dent*, **101**, 405–412.

20 Beresin, V. E., and Schiesser, F. J. (1979) *Neutral Zone in Complete and Partial Dentures*, 2nd edn. Mosby, St. Louis, MO, pp. 15, 73–108, 158–183.

10

Denture Placement

Introduction

Following the trial placement appointment, the wax dentures are contoured appropriately, sealed to the definitive cast, and sent to the laboratory for fabrication along with detailed written instructions [1]. The dentures may be processed by a compression-molding technique [1], an injection-molding technique [2], or they may be digitally milled [3, 4]. The denture-processing procedure should minimize tooth movement or an inadvertent increase in the occlusal vertical dimension (OVD) [1]. While processing the dentures, it is important to use the technique and materials that lead to minimum distortion [1]. Computer-aided design and computer-aided manufacturing (CAD/CAM) milling of dentures produces a better fit of the denture base and less movement of the denture teeth than conventional processing techniques [5].

The final clinical appointment in the denture-fabrication process is the denture-placement appointment. It is a critical appointment. Prior treatment procedures and materials can incorporate several inaccuracies into the completed prostheses despite meticulous care and good technique. Denture placement should be considered as an appointment for eliminating errors introduced in the prostheses [6]. It is important that the dental practitioner allocate sufficient time for this appointment to evaluate and adjust the new prosthesis carefully, educate patients (regarding the physiologic adjustments, maintenance and home care), and address all their concerns and questions, thereby aiding the adaptation and acceptance of the new prosthesis [6]. This chapter outlines a step-by-step protocol to be followed during the denture-placement appointment.

Before placing the dentures in the patient's mouth it is imperative that the borders, intaglio, and the cameo surface be thoroughly evaluated both by touch and with magnifying loupes for scratches, sharp spicules, and imperfections (Figures 10.1 and 10.2). All the borders should be rounded and all imperfections and sharp spicules should be eliminated before trying the dentures in the patient's mouth.

Placement of Immediate Dentures

For patients receiving immediate dentures, the remaining natural teeth are removed (as atraumatically as possible) at this appointment. A surgical template (fabricated by the laboratory or the clinician) may be used to guide the contouring of alveolar bone, when indicated. After adjusting all surfaces (described below), the dentures are placed in the mouth and the patient is instructed not to remove them and to return for a recall appointment the next day. At the 24-hour recall appointment the dentures are removed by the clinician, evaluated, and adjusted accordingly.

Application of the Neutral Zone in Prosthodontics, First Edition. Joseph J. Massad, David R. Cagna, Charles J. Goodacre, Russell A. Wicks and Swati A. Ahuja.
© 2017 John Wiley & Sons, Inc. Published 2017 by John Wiley & Sons, Inc.
Companion website: www.wiley.com/go/massad/neutral

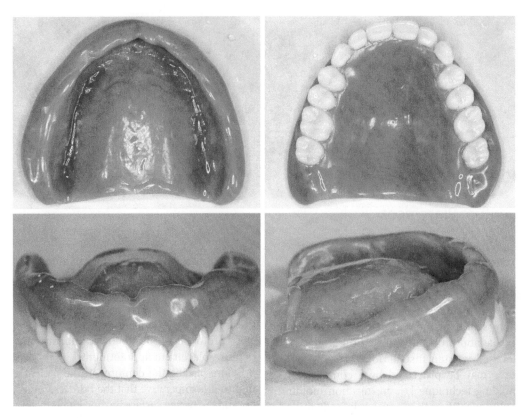

Figure 10.1 Definitive maxillary denture. Intaglio surface (top left), occlusal surface (top right), Polished surface (frontal view, bottom left), Polished surface (lateral view, bottom right).

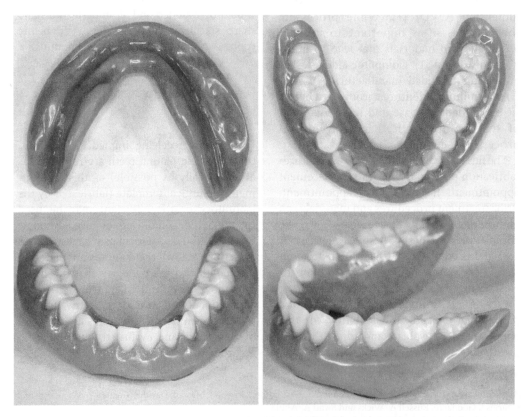

Figure 10.2 Definitive mandibular denture. Intaglio surface (top left), Occlusal surface (top right), Polished surface (frontal view, bottom left), Polished surface (lateral view, bottom right).

Placement Procedures

The following procedures are carried out during the denture-placement appointment:

- evaluation and adjustment of intaglio surface, denture borders, and cameo surface;
- evaluation and refinement of occlusion; and
- postplacement patient instructions.

Evaluation and Adjustment of Intaglio Surface

Complete dentures should exhibit uniform and positive tissue contact without applying excessive pressure on the denture-bearing tissues. Excessive pressure may lead to inflammation, soreness, patient discomfort, and ultimately treatment failure. The intaglio surface of the dentures should be disclosed to identify the areas exhibiting excessive tissue pressure, and appropriately relieved [7]. Certain clinical situations/conditions may lead to excessive tissue pressure, including: (i) residual ridge undercuts relative to the path of prosthesis placement; (ii) excessive tissue displacement during the making of definitive impressions [8]; (iii) bony ridges or prominences covered by thin, nonyielding denture-bearing tissue; (iv) processing shrinkage [9, 10]/distortion or (v) soft and hard-tissue changes [11–13].

It is crucial to evaluate and adjust complete dentures one at a time to optimize fit and identify and relieve excessive denture base pressure. The dentures should be adapted to the edentulous ridges before refining the occlusion. Pressure-indicating paste (PIP) or pressure-disclosing paste is considered as the material of choice for identifying and locating denture base interferences to complete placement of the prostheses [14, 15]. The use of PIP is easy, simple, and quick. It is also economical and yields consistent and accurate results [14, 15]. It is recommended during denture placement and all postplacement visits where adjustment of denture-base tissue contact would be indicated.

Technique

Evaluation of the intaglio surface of the dentures should be done, one denture at a time. The intaglio surface of the denture is dried and a stiff brush is used to apply PIP onto the intaglio surface of the denture base (in a thin layer), using unidirectional brush strokes. PIP is applied to only half of the intaglio surface to prevent the appearance of erroneous excessive contact areas during insertion of the denture (Figure 10.3a). To avoid paste displacement due to contact with lips, tongue, or ridge prominence, the denture, with PIP applied, is immersed in a bowl of water at room temperature (Figure 10.3b) and then carefully inserted into the patient's mouth using a rotational path of insertion (Figure 10.3c). Once accurately seated over the edentulous ridge, firm pressure (but not excessive pressure) is applied with fingers for a few seconds. The denture is removed gently and the paste distribution is observed. The PIP paste distribution is interpreted as follows [15]: (i) undisturbed brush-stroke pattern suggests no tissue contact; (ii) evenly disrupted stroke pattern but no visible denture base suggests desirable tissue contact; (iii) displacement of paste with denture base visible, or "showing through," suggests substantial or excessive tissue contact (Figure 10.4).

Before any adjustments are made to the denture it is important to repeat the procedure to verify and ensure that the same areas are being disclosed. The excessive contact areas are relieved using an appropriately shaped acrylic resin bur or foam wheel (Figure 10.5) [16]. The residual acrylic resin shavings and the PIP are removed with a gauze pad and the procedure is repeated until an even layer of PIP with disrupted stroke pattern and no visible denture base is seen on the intaglio surface of the denture [15]. This procedure should be performed carefully to avoid excessive adjustments on the denture base, which could lead to loss of prosthesis retention, support, and stability, and also increase fracture susceptibility. The same technique should be utilized for both dentures to disclose excessive pressure areas and improve fit.

(a) (b)

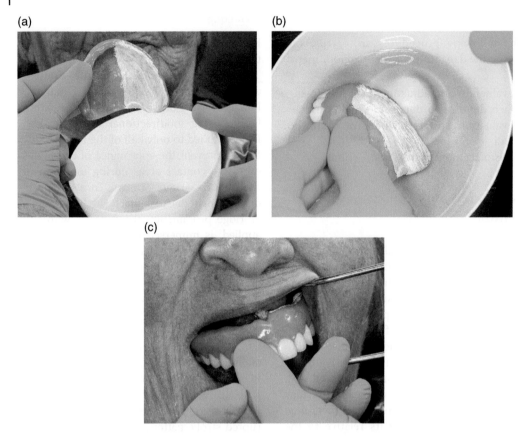

(c)

Figure 10.3 (a) PIP applied on intaglio surface of denture base (in a thin layer) using unidirectional brush strokes; (b) denture base with PIP applied, is immersed in a bowl of room temperature water; (c) Denture carefully inserted in the patient's mouth with an optimal path of insertion.

Figure 10.4 Displacement of paste with denture base visible, suggests areas of substantial or excessive tissue contact.

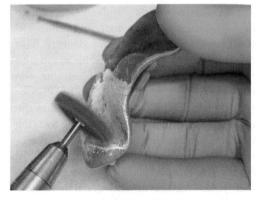

Figure 10.5 Reticulated foam polishing wheel used to relieve and refine excessive contact areas.

Evaluation and Adjustment of Denture Borders

Following the evaluation and adjustment of the intaglio surface, the next step is to assess the denture borders. Borders of a complete denture should be in harmony with the anatomic and functional limits of the denture foundation and adjacent tissues. The appropriate length and dimension of denture borders is critical to establish a peripheral seal. Underextended borders may compromise the peripheral seal, retention, and stability of dentures. Overextended denture borders may encroach on muscle attachments leading to soreness, ulceration of the tissues, and denture dislodgement during function. Definitive impressions of severely resorbed ridges may be overextended. Accordingly, the prostheses made from such impressions would need appropriate adjustment at the time of denture placement [17, 18].

While evaluating maxillary denture fit, critical evaluation of disclosing paste patterns along the posterior denture border is indicated. If the paste remains undisturbed in this area, a lack of tissue contact is evident. Additive correction of the posterior denture border may be required to develop a favorable postpalatal seal. The procedure includes applying a suitable bonding agent on the intaglio and cameo surface of the posterior denture border. Then a 5 mm-wide strip of light polymerizing resin is adapted onto the posterior extent of the intaglio surface and wrapped over the posterior denture border onto the cameo surface. The denture is placed in the mouth and seated firmly. To extend the PPS area physiologically, the patient may be instructed to sip and swallow water several times. The patient should also occlude the nostrils and attempt to blow out through the nose. This is known as the Valsalva maneuver; it causes the soft palate to drape downward along the posterior denture border and mold the unpolymerized resin. Once the resin has been molded physiologically, a curing light may be used intraorally to initiate polymerization of the resin. The denture is removed from the mouth and inspected to evaluate the adaptation of the resin to the denture. It is placed in a laboratory light curing unit to complete the polymerization of the resin. The newly added resin is trimmed, finished and polished for final placement.

Disclosing wax or PIP have also been utilized for the evaluation of denture borders [6, 19–21]. The addition of petrolatum to disclosing wax helps decrease its viscosity and makes it easier to handle and manipulate. This mixture can be loaded in to plastic syringe for neat and easy dispensing onto the denture border to be evaluated. Fast-polymerizing vinyl polysiloxane (VPS) interocclusal registration material can also be used to evaluate denture borders [21]. It is easier to clean away than PIP but slightly more expensive [21]. The technique for assessing denture borders is described below.

Borders should also be evaluated one arch at a time. The VPS-disclosing material is dispensed on a border of the denture (Figure 10.6a). The denture is carefully seated in the mouth avoiding displacement of the recording media due to contact with the lips or tongue, or ridge prominence. While the denture is held accurately in place from the occlusal surface, the patient is instructed to perform appropriate border-molding movements (described in detail in Chapter 3) (Figure 10.6b). The denture is removed from the mouth gently following the polymerization of the VPS-disclosing material and its distribution is observed (Figure 10.6c). Displacement of the VPS material with denture base visible suggests substantial or excessive tissue contact (overextended border). An indelible marker is painted over the disclosing material and the visible portion of the denture border and the VPS material is carefully removed (Figure 10.6d) [21]. The areas of "show through" or overextension (marked areas) (Figure 10.6e) are adjusted using an appropriately shaped resin acrylic bur or foam wheel on a slow-speed laboratory rotary hand piece (Figure 10.6f). The procedure is

(a)

(b)

(c)

(d)

(e)

(f)

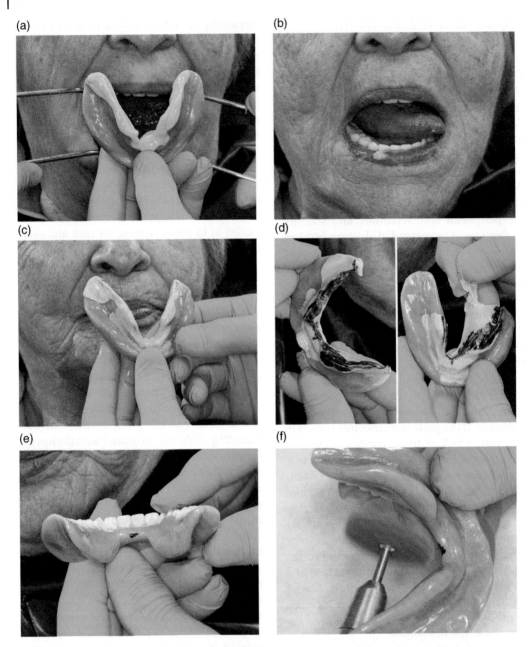

Figure 10.6 (a) VPS material dispensed on the lingual borders of the mandibular denture; (b) patient performs tongue movements to mold the VPS material physiologically; (c) denture with the VPS material removed from the patient's mouth and analyzed for excessive pressure; (d) indelible marker painted over the disclosing material and visible portion of denture border (left). VPS material removed from the denture border (right); (e) The marked area represents the overextended area; (f) Areas of overextension (marked areas) adjusted using reticulated foam polishing wheel. Note: the reticulated foam wheel removes the marked areas while leaving a smooth surface.

repeated until no visible denture base is seen on the border of the denture. The procedure is repeated for all the borders of the dentures. This procedure should be performed carefully to avoid excessive removal of denture border, which could lead to loss of retention. After trimming the borders, they should be appropriately finished and polished [22].

Evaluating the Cameo Surface

The polished / cameo surface / external surface should not be developed arbitrarily but by making an external impression of the polished surface of the wax trial denture and carrying the form through construction of the definitive prostheses (see Chapter 9) [23–25]. Accurately developed polished surfaces not only help with the retention and stability of complete dentures but also aid in improving speech, mastication, and increasing patient comfort and adaptability [23]. Inappropriately contoured polished surfaces may result in a loose and unstable denture [23–25]. It is critical to evaluate the polished surfaces as meticulously as the impression and the occlusal surfaces [6].

Evaluation of the cameo surface should also be completed one arch at a time. The cameo surface of the denture is evaluated at the time of denture placement as follows: fast-polymerizing VPS interocclusal registration material is thinly applied on the buccal cameo surface of the maxillary denture base (Figure 10.7a). The denture is placed in the patient's mouth carefully to avoid material displacement due to contact with lips, tongue, or ridge prominence (Figure 10.7b). The patient is instructed to perform the sequence of oral movements as described in Chapter 8 (Figure 10.7c). The denture is carefully removed and the disclosing material distribution is observed on the buccal cameo surface of the denture (Figure 10.7d). An indelible marker is painted over the disclosing material and the visible portion of the

denture border (Figure 10.7e). The VPS material is carefully removed and the areas of show through or excessive contact (marked areas) (Figure 10.7f) are adjusted using an appropriately shaped resin acrylic bur or foam wheel on a slow speed laboratory rotary handpiece (Figure 10.7g). The procedure is repeated until no visible denture base is seen. This procedure should be performed carefully to avoid excessive removal of denture base, which could lead to loss of retention and make the denture susceptible to fracture. The same procedure is repeated for the lingual / palatal surfaces and followed with the mandibular denture. After adjusting the denture cameo surface, it should be appropriately finished and polished [22]. All the surfaces should be smooth and free of scratches and imperfections. The denture base should have a uniform thickness of 2.0–2.5 mm so that it does not feel bulky and uncomfortable to the patient.

Occlusal Evaluation and Correction

Stable and retentive denture bases are a prerequisite to properly equilibrate the occlusion. Hence, occlusal equilibration is performed following the evaluation and adjustment of the intaglio surface, borders, and the cameo surface. Occlusal inaccuracies in dentures may result from tissue changes after impressions, differences in fit of the record base and the denture base on the supporting tissues, inaccurate jaw relationship records, inaccurate mounting of the casts on the articulator, and processing procedures [26–29]. Occlusal disharmony may lead to soreness, patient discomfort, difficulty in mastication, denture instability, and damage to the denture-bearing tissues [27]. A single point of occlusal prematurity has the potential to disrupt the entire denture occlusion and prevent proper articulation [28].

There are multiple methods of refining the occlusion with complete dentures. They include obtaining clinical records that

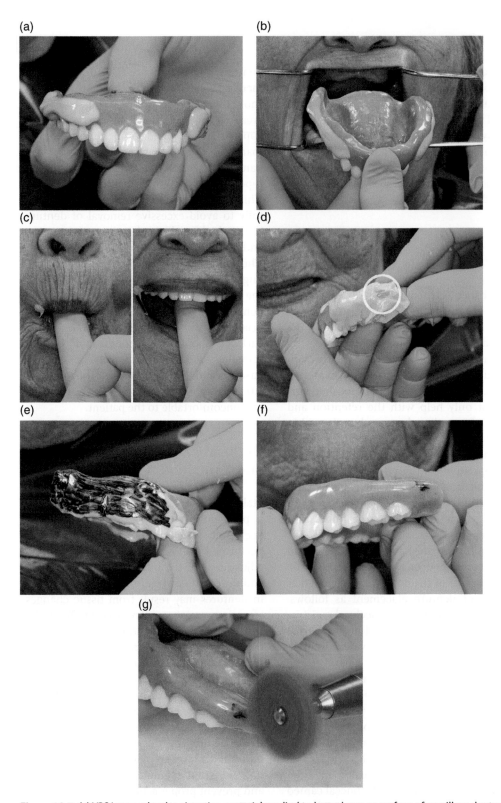

Figure 10.7 (a) VPS interocclusal registration material applied to buccal cameo surface of maxillary denture; (b) Cheeks retracted and the denture is carefully placed in the patient's mouth; (c) patient is asked to perform various movements to physiologically mold the VPS material; patient is instructed to suck on the finger (left); patient is instructed to smile broadly (right); (d) denture is removed from the patient's mouth and areas of show through are evaluated; (e) indelible marker painted over disclosing material and visible portion of denture surface; (f) the marked area represents the overextended area; (g) areas of excessive contact (marked areas) adjusted using reticulated foam polishing wheel.

permit adjustment on an articulator (clinical remount technique) as well as intraoral methods [27, 28]. Because the resiliency and varying compressibility of the denture bearing tissues leads to displacement of the denture bases and may generate multiple erroneous markings, some clinicians prefer the clinical remount technique [27, 30]. Some also lend preference to this technique based on the magnitude of the observed discrepancy. This procedure requires remount casts, interocclusal records, mounting the remount casts in the articulator using interocclusal records, verifying the mounting, and adjusting the occlusion of the dentures on the articulator [26, 30]. While this procedure is effective, it is also time consuming, and requires a modest amount of laboratory procedures.

A particularly effective intraoral method is the use of the jaw recorder device (Intraoral Gothic arch tracer). The use of this device for registering the maxillo-mandibular jaw relationship record is described in Chapter 6. The jaw recorder device can also be used to refine the occlusion of the definitive prostheses [31, 32]. It stabilizes the maxillary and the mandibular dentures and aids in the establishment and recording of consistent, repeatable, and accurate OVD and CR, making it possible to identify true occlusal interferences and equilibrate the occlusion [31, 33]. Conventional gothic arch tracers are technique sensitive, complex, and difficult to use but the new central bearing point device is simple, easy to use, and cost and time effective. It can be used with both the subtractive and the additive techniques of occlusal equilibration:

- Subtractive technique – grinding prosthetic tooth surfaces in order to refine occlusal contacts. This is by far the most common approach to occlusal adjustment.
- Additive technique – this method involves adding light polymerized composite resin to occlusal voids in the mandibular posterior prosthetic teeth and allowing the patient to form appropriate occlusal morphology.

Subtractive Correction Technique

The intraoral jaw recording device is mounted on the complete dentures in the same way it is mounted on the record bases as described in Chapter 6 (Figure 10.8). The dentures are placed in the mouth and the threaded pin is adjusted so that the occlusal surfaces of the prosthetic teeth are slightly out of contact. A thin articulating film / ribbon / strip is placed between the prosthetic teeth to verify the lack of contacts (Figure 10.9). Gradually the pin height is reduced until the introduction of articulating film identifies the first occlusal

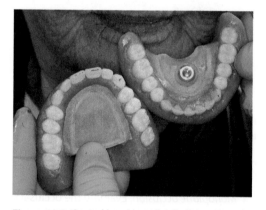

Figure 10.8 Central bearing point device mounted on complete dentures to diagnose occlusal discrepancies.

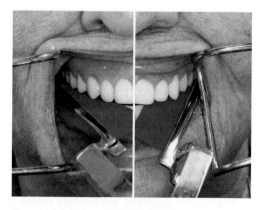

Figure 10.9 Thin articulating film placed between prosthetic teeth to mark first premature contacts.

contact(s) when the patient taps up and down in CR position. The premature contacts are adjusted using appropriate rotary instruments (Figure 10.10). The centric pin height is gradually reduced to identify subsequent occlusal interferences and appropriate corrections are accomplished until bilateral simultaneous centric posterior contacts are demonstrated. It is important to perform methodical selective grinding to aid in

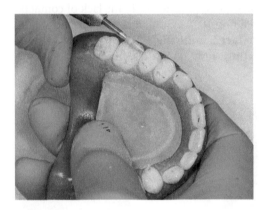

Figure 10.10 Premature contacts adjusted using appropriate rotary instruments.

maintaining the tooth form and also in developing a harmonious interference free occlusion. Care must also be executed during this procedure so as to not change / decrease the established OVD in the definitive prosthesis. Occlusal correction specifications for various occlusal schemes [34, 35] are described in Box 10.1.

Additive Correction Technique

For this procedure, large class-I cavity preparations (with slight undercuts) are made in all mandibular posterior prosthetic teeth (Figure 10.11a) and the intraoral jaw recording device is mounted on the prosthesis. The dentures are seated in the mouth and the central bearing (threaded) pin is raised to the planned occlusal vertical dimension. In this position, no posterior occlusal contacts should be present due to the class-I cavity preparations in mandibular prosthetic teeth. The mandibular denture is removed from the mouth. A suitable bonding agent is applied as appropriate to the prepared teeth surfaces and light polymerized composite resin is

---

Box 10.1 Occlusal correction specification for various occlusal schemes

- *Monoplane occlusion.* Evaluate the occlusal surfaces of the maxillary and mandibular posterior teeth against a flat plane and correct them as necessary. Occlusal interferences should be identified intraorally and eliminated without disrupting the flat occlusal plane.
- *Fully anatomic balanced occlusion.* The occlusion should be evaluated and adjusted in centric relation position first and then in eccentric positions. Instead of altering the stamp cusps (to maintain their shape, size, and form) the opposing fossa should be adjusted / reshaped and made deeper or wider. All cusps should glide smoothly through the grooves and embrasures of opposing dentitions. When evaluating eccentric contacts, the use of a different-color articulating ribbon might be helpful to distinguish between the eccentric contacts and the previously marked centric contacts. Subtle eccentric mandibular movement should be accomplished during the course of this procedure and appropriate occlusal adjustments made to permit smooth occlusal function in both centric and eccentric mandibular positions.
- *Lingualized occlusion.* It is important to ensure that the maxillary buccal cusps do not contact the opposing dentition. The palatal / lingual cusps of the maxillary prosthetic teeth should be the only contact with mandibular opposing prosthetic teeth. The reduction of the maxillary lingual cusps should be avoided and instead the mandibular prosthetic teeth should be adjusted to achieve an interference free harmonious occlusion. For lingualized balanced occlusion, the occlusion should be evaluated and adjusted in centric relation position first and then in eccentric positions. Appropriate occlusal adjustment should be made to permit smooth occlusal function in both centric and eccentric mandibular positions.

(a)

(b)

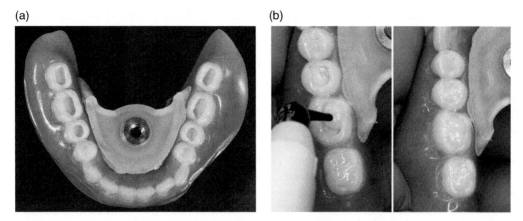

Figure 10.11 (a) Large class I cavity preparations made in all mandibular posterior prosthetic teeth; (b) Light polymerized composite resin filled (left) and carefully condensed in to the preparation to avoid overfilling (right).

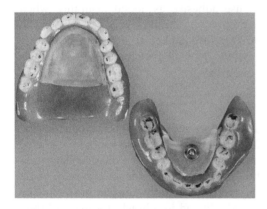

Figure 10.12 Occlusal contacts verified with articulating paper. Note: this procedure is performed for both the additive and the subtractive technique.

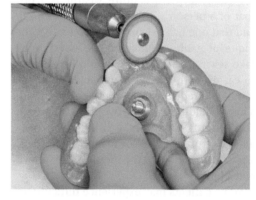

Figure 10.13 Occlusal surfaces finished and polished in the laboratory.

carefully condensed in to the preparation to avoid overfilling (Figure 10.11b) [36]. A very thin layer of a lubricant such as petrolatum may be applied to the maxillary occlusal surfaces to prevent the resin from adhering to them [36]. The mandibular denture is carefully seated in the mouth and the patient is instructed to make various mandibular movements (prostrusion, retrusion, left lateral, and right lateral). This procedure aids in the development of functionally generated occlusal pathways [36]. The mandibular denture is removed from the mouth and the occlusal surfaces of the prosthetic teeth are inspected for voids and or excess resin. If necessary, more material may be added or

excess may be trimmed with a handheld composite resin instrument. A laboratory light curing unit may be used to polymerize the resin. Occlusal contacts are verified again with articulating paper (Figure 10.12) and the occlusal surfaces are finished and polished appropriately (Figure 10.13) [37].

Patient Education and Instructions

During the diagnosis and treatment planning appointment, patients should be informed regarding the limitations of complete dentures and these should be restated at the time of denture placement. They also should be reminded to not compare themselves with other people (wearing dentures) as every patient is different and also the adaptive capacity of each patient varies depending on their age, general health, personality, and emotional state [34, 35]. Patients should be informed that a minimum 3–4-week period is required to learn to use new dentures and some patients may require even more time.

Patients should be made to understand that new dentures may feel strange and give them a sense of fullness of the oral cavity including the lips and cheeks but with time they will adjust and accommodate to them. Patients should also be informed that chewing efficiency of a denture wearer is 1/6th that of a person with natural dentition [38], and that one has to learn to master the technique of eating with dentures, which may take several weeks. They should be advised to eat on both the sides and start with soft food and gradually progress to coarser food. They should be asked to avoid incising with their front teeth, as it may dislodge the maxillary denture causing social embarrassment, and perform most of the incising with a fork and a knife.

Some patients may experience difficulty in speaking with new dentures [39]. These patients should be instructed to practice speech diligently with new dentures by reading aloud difficult words, sentences or phrases while enunciating slowly and deliberately [39]. Patients should be comforted and informed that their speech will improve in a few weeks. Increased salivation may be associated with the introduction of new dentures (foreign object) in the mouth and will gradually decrease with time. This increased salivation may impair mastication and speech. Patients should be advised to swallow their saliva instead of continuously spitting it out. Soreness is a common occurrence with new dentures and patients should be instructed to discontinue wearing dentures and schedule an appointment with the dental practitioner for denture adjustment

when they develop sore spots [16]. They should be instructed that sore spots will not fade away unless the denture is relieved in the pressure areas. They should be asked to wear their dentures a day prior to their visit so that the sore spots are easily visible and permit easy and accurate adjustments to the denture base.

It is also important to discuss adhesives with the patients and assure them using small amounts of adhesives is perfectly acceptable as long as they practice regular hygiene maintenance. Demonstrating proper use versus improper use and how to remove adhesive adequately from the prostheses and oral tissues will give the patient the confidence to use adhesive as added insurance when appropriate. Adhesives work best when small amounts are applied to well-fitting prostheses [40]. Not all patients will need to use adhesives.

Technique for Adhesive Application

The denture should be dried completely and 3–4 small beads of adhesive should be placed on the intaglio surface of the maxillary denture and spread to the entire ridge area. (Figure 10.14a and b). The denture should be quickly dipped in a bowl of water and then inserted in the mouth.

Technique for Adhesive Removal

Adhesive can be easily removed by placing the prosthesis in a bowl of warm water for 30 seconds. Then, a mechanical toothbrush is inserted in the water bowl and used to clean the adhesive off the denture. The mechanical toothbrush can also be used to remove the adhesive effectively from the denture-bearing tissues.

Only a small quantity of adhesive should be used on dentures. The need for the use of excess quantities of adhesive for denture retention indicates that the dentures do not fit and may need to be remade or relined. Excess adhesives may alter the OVD, the existing occlusion and stability of the denture.

(a)

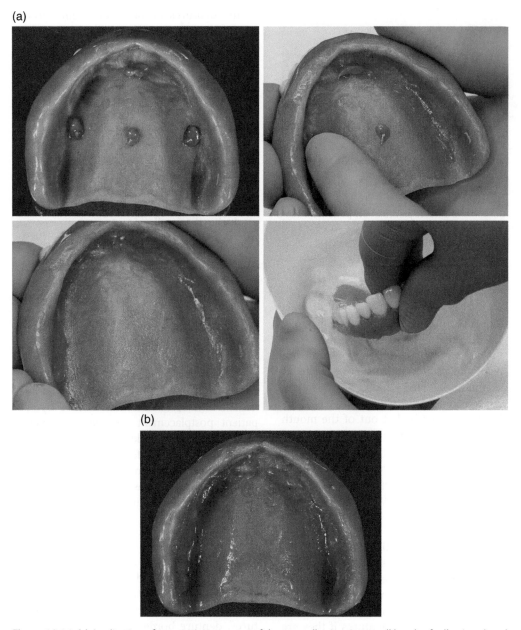

(b)

Figure 10.14 (a) Application of appropriate amount of denture adhesive: 3–4 small beads of adhesive placed on the intaglio surface of the maxillary denture (top left) and spread to the entire ridge area (top right); Adhesive evenly spread (bottom left) and the denture is quickly dipped in a bowl of water (bottom right); (b) adhesive appropriately applied to maxillary denture.

At the time of denture placement (Figure 10.15), the patient should also be informed in detail regarding the expected maintenance cost and replacement frequency (on average every 5–7 years) in order to maintain a good and long-term patient-provider relationship. Patients should be given a written copy of denture home-care instructions [34, 35] to aid them in maintaining their prostheses and oral health.

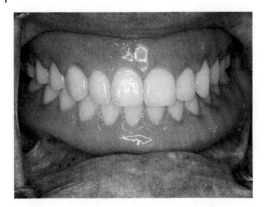

Figure 10.15 Dentures placed in the patient's mouth.

Home Care Instructions for Denture Patients

1) To maintain healthy mucosal tissue, moisten the oral cavity, and massage and clean the mucosa, tongue, and roof of the mouth daily with a soft toothbrush (or a mechanical toothbrush) for 5 minutes in the morning and 5 minutes in the evening.

2) Dentures *must* be left out of the mouth for at least 7–8 hours in a 24-hour period to provide the necessary rest to the denture-bearing tissues.

3) A stiff denture-cleaning brush and diluted dish soap solution should be used for cleaning the dentures a minimum of two times per day. Never use toothpaste or mouth rinse for cleaning the dentures as they may abrade or stain the dentures. Commercially produced denture cleaning effervescent tablets may be used as an additional aid for cleaning the dentures.

4) To prevent breaking dentures if they are accidentally dropped, brush the dentures over a towel, a soft mat, or a sink filled with water.

5) When the dentures are left out of the mouth, store the dentures in a closed container filled with water to prevent dehydration and distortion of dentures. Rinse well in the morning before reinserting.

6) Tissues that support the denture are constantly changing [13, 41], which could result in loosening of the denture. When the dentures become consistently loose they need to be refitted or replaced, and an appointment is required with the dental practitioner for evaluation.

7) Instruct the patient not to adjust or reline the dentures.

8) An annual maintenance visit is necessary to evaluate the supporting tissues and prostheses. Typically, the denture placement appointment will be followed by a postplacement appointment after 24 hours, then one week, and again at one month. The same procedures will be accomplished during the recall appointments.

Postplacement Problems with New Dentures

Pain thresholds vary drastically among patients. Some patients may complain of every minor discomfort whereas some may have serious problems and still not complain at all; hence regular maintenance visits are critical. Most of the problems encountered by the patient postplacement are minor problems and can be solved quickly and easily, but when they persist unnoticed for a long period of time they can surface as a major problem. Some of the common problems that can occur with complete dentures include the following.

Retention Problems

Retention problems are most commonly caused by overextended or underextended denture borders and/or thick denture borders. When a patient complains of loose dentures, denture borders should be evaluated and adjusted (as described in the previous section). Retention problems may also be associated with a dry mouth, heavy secretion of mucinous saliva, retracted tongue position and/or lack of neuromuscular coordination. All these conditions should be diagnosed prior to the initiation of definitive therapy and the patient should be informed regarding their impact on the prognosis of the dentures, and if treatment can be altered to improve the prognosis.

Occlusal interferences may also affect the retention of the prostheses they should be identified and eliminated.

Soreness

Excessive pressure on areas with thin mucosa (by the prostheses) could lead to soreness. The intaglio surface of the denture should be adequately disclosed and relieved appropriately. Overextended, thick, or sharp borders may also lead to soreness and should be disclosed and relieved accurately. Occlusal interferences may also lead to soreness; they should be identified and eliminated (Figure 10.16). In addition, inadequate interocclusal distance can cause soreness, sometimes necessitating remake of one of the dentures at the optimal OVD.

Clenching, bruxism, nutritional deficiencies, and conditions such as diabetes or pemphigus vulgaris may also affect tissues, leading to soreness.

Sore Throat

Overextension in the region of soft palate, hamular notch, distobuccal aspect of maxillary denture (Figure 10.17) or distolingual aspect of mandibular denture or the pterygomandibular raphe area may lead a sore throat. All the areas of overextension should be examined and relieved appropriately.

Speech Problems

Patients with a history of corrected speech problems will usually experience speech problems with new prostheses. They may be referred to a speech therapist for examination and treatment.

Speech problems can also be caused due to an excessive increase in OVD or excessive horizontal and/or vertical overlap between the anterior teeth. In either of these cases the prosthesis will need to be remade at the correct OVD.

Gagging

Overextended and thick denture borders can lead to patient gagging with the new prostheses (Figure 10.18). Overextended and thick borders should be disclosed and adjusted appropriately. Improper positioning of the occlusal plane, setting the mandibular posterior teeth too far lingually

Figure 10.17 Thick left distobuccal border of the maxillary denture may cause sore throat.

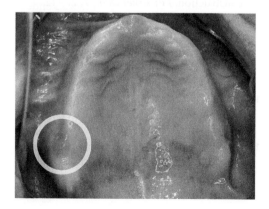

Figure 10.16 Occlusal interferences may cause sore spots or ulcerations on the ridges.

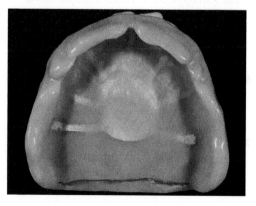

Figure 10.18 Thick posterior border of the maxillary denture may lead to gagging.

and / or excessive increase in OVD can all lead to patient gagging, necessitating remake of the dentures.

Summary

Dental practitioners should constantly assure patients that they are available should any concerns arise. When placing a new prosthesis, it is vital to evaluate and adjust it to aid the process of patient adaptation and acceptance of the new prosthesis. It is also critical to educate and train the patient regarding proper care of dentures to maintain optimal service. Patients may be given several samples to assist them in cleaning and maintaining their dentures such as effervescent tablets, a denture-cleaning toothbrush, and a denture case. It is also important to explain and give the patient a copy of written instructions to maintain the prosthesis and their oral health.

References

1 Rudd, K. D. (1964) Processing complete dentures without tooth movement. *Dent Clin North Am*, **8**, 675–691.

2 Pryor, W. J. (1942) Injection molding of plastics for dentures. *J Am Dent Assoc*, **29**, 1400–1408.

3 Kattadiyil, M. T., Goodacre, C. J., and Baba, N. Z. (2013) CAD/CAM complete dentures: a review of two commercial fabrication systems. *J Calif Dent Assoc*, **41**, 407–416.

4 Goodacre, C. J., Garbacea, A., Naylor, W. P., *et al.* (2012) CAD/CAM fabricated complete dentures: concepts and clinical methods of obtaining required morphological data. *J Prosthet Dent*, **107**, 34–46.

5 Goodacre, B. J., Goodacre, C. J., Baba, N. Z., and Kattadiyil, M. T. (2016) Comparison of denture base adaptation between CAD/CAM and conventional fabricationtechniques. *J Prosthet Dent*. doi: 10.1016/j.prosdent.2016.02.017

6 Keubker, W. A. (1984) Denture problems. Causes, diagnostic procedures and clinical treatment. I. retention problems. *Quintessence Int*, **15**, 1031–1044.

7 Jankelson, B. (1962) Adjustment of dentures at time of insertion and alterations to compensate for tissue change. *J Am Dent Assoc*, **64**, 521–531.

8 Woelfel, J. B. (1962) Contour variations in impression of one edentulous patient. *J Prosthet Dent*, **12**, 229–254.

9 Woelfel, J. B., Paffenbarger, G. C., and Sweeney, W. T. (1960) Dimensional changes occurring in dentures during processing. *JADA*, **61**, 413–430.

10 Phoenix, R. D. (1996) Denture base resins: Technical considerations and processing techniques. In: *Phillip's science of dental materials*, 10th Ed Philadelphia, WB Saunders Company, pp. 237–271.

11 Chase, W. W. (1961) Tissue conditioning utilizing dynamic adaptive stress. *J Prosthet Dent*, **11**, 804–815.

12 Lytle, R. B. (1957) The management of abused oral tissues in complete denture construction. *J Prosthet Dent*, **7**, 27–42.

13 Tallgren, A. (2003) The continuing reduction of the residual alveolar ridges in complete denture wearers: a mixed-longitudinal study covering 25 years. *J Prosthet Dent*, **89**, 427–435.

14 Gronas, D. G. (1977) Preparation of pressure indicating paste. *J Prosthet Dent*, **37**, 92–94.

15 Stevenson-Moore, P., Daly, C. H., and Smith, D. E. (1979) Indicator pates: Their behavior and use. *J Prosthet Dent*, **41**, 258–265.

16 Keubker, W. A. (1984) Denture problems. Causes, diagnostic procedures and clinical treatment. I I. patient discomfort problems. *Quintessence Int*, **15**, 1131–1141.

17 Hickey, J. C., Zarb, G. A., and Bolender, C. L. (1985) *Boucher's Prosthodontic Treatment for Edentulous Patients*, Mosby, St Louis, MO, p. 41.

18 Clancy, J. M. (1988) A technique for limiting reduction of overextended denture borders. *J Prosthet Dent*, **60**, 258–259.

19 Phoenix, R. D., and DeFreest, C. F. (1997) An effective technique for denture border evaluation. *J Prosthodont*, **6**, 215–217.

20 Logan, G. I., and Nimmo, A. (1984) The use of disclosing wax to evaluate denture extensions. *J Prosthet Dent*, **51**, 280–281.

21 Haeberle, C. B., Abreu, A., and Metzler, K. (2015) Use of a bite registration vinyl polysiloxane material to identify denture flange overextension and / or excessive border thickness in removable prosthodontics. *Gen Dent*, **63**, e32–35.

22 Kuhar, M., and Funduk, N. (2005) Effects of polishing techniques on the surface roughness of acrylic denture base resins. *J Prosthet Dent*, **93**, 76–85.

23 Fish, E. W. (1937) *Principles of Full Denture Prosthesis*, John Bale Medical Publications, London.

24 Beresin, V. E., and Schiesser, F. J. (1976) The neutral zone in complete dentures. *J Prosthet Dent*, **95**, 93–100.

25 Raybin, N. H. (1963) The polished surface of complete dentures. *J Prosthet Dent*, **13**, 236–239.

26 Nandal, S., Himanshu, S., and Ghalaut, P. (2014) Complete-denture insertion appointment: What to look for? *IJARESM*, **2**, 4–13.

27 Shigli, K., Angadi, G. S., and Hegde, P. (2008) The effect of remount procedures on patient comfort for complete denture treatment. *J Prosthet Dent*, **99**, 66–72.

28 Polyzois, G. L., Karkazis, H. C., and Zissis, A. J. (1991) Remounting procedures for complete dentures: a study of occlusal contacts by the photocclusion technique. *Quintessence Int*, **22**, 811–815.

29 Vig, R. G. (1975) Method of reducing the shifting of teeth in denture processing. *J Prosthet Dent*, **33**, 80–84.

30 Ansari, I. H. (1996) Simplified clinical remount for complete dentures. *J Prosthet Dent*, **76**, 321–324.

31 Massad, J. J., and Connelly, M. E. (2000) A simplified approach to optimizing denture stability with lingualized occlusion. *Compend Contin Educ Dent*, **21**, 555–558, 560, 562.

32 Young, L. Jr, and Johnson, C. (1987) Adjusting complete denture occlusion with an intraoral balancer. *Compendium*, **8**, 54–56, 58.

33 Utz, K. H., Müller, F., Bernard, N., *et al.* (1995) Comparative studies on check-bite and central-bearing-point method for the remounting of complete dentures. *J Oral Rehabil*, **22**, 717–726.

34 Plummer, K. D. (2009) Insertion, in *Textbook of Complete Dentures*, 6th edn. (eds. A. O. Rahn, J. R. Ivanhoe, and K. D. Plummer). People's Medical Publishing House, Shelton, CT, 229–249.

35 Chang, T.-L., and Fenton, A. H. (2012) Prosthesis insertion and follow-up appointments, in *Prosthodontic Treatment for Edentulous Patients: Complete Dentures and Implant-Supported Prostheses*, 13th edn (eds. G. Zarb, J. A. Hobkirk, S. E. Eckert, and R. F. Jacob). Elsevier Inc., St. Louis, MO, pp. 255–280.

36 Ruffino, A. R. (1984) Improved occlusal equilibration of complete dentures by augmenting occlusal anatomy of acrylic resin denture teeth. *J Prosthet Dent*, **52**, 300–302.

37 Kuhar, M., and Funduk, N. (2005) Effects of polishing techniques on the surface roughness of acrylic denture base resins. *J Prosthet Dent*, **93**, 76–85.

38 Kapur, K. K., and Soman, S. D. (2006) Masticatory performance and efficiency in denture wearers. *J Prosthet Dent*, **95**, 407–411.

39 Keubker, W. A. (1984) Denture problems. Causes, diagnostic procedures and clinical treatment. III/Iv. Gagging problems and speech problems. *Quintessence Int*, **15**, 1231–1238.

40 Kapur, K. K. (1967) A clinical evaluation of denture adhesives. *J Prosthet Dent*, **18**, 550–558.

41 Atwood, D. A. (1971) Reduction of residual ridges: A major oral disease entity. *J Prosthet Dent*, **26**, 266–279.

11

Use of CAD/CAM Technology for Recording and Fabricating Neutral-Zone Dentures

Introduction

Computer-aided design / computer-aided manufacturing (CAD/CAM) technology has been used in dentistry since the early 1980s. Andersson envisioned the use of titanium for crown fabrication and pioneered the CAD/CAM fabrication process, which resulted in cementation of the first CAM fabricated titanium crown in 1982 [1]. Mörmann developed a prototype CAD/CAM system in 1983 and placed the first chairside fabricated ceramic restoration in 1985 [2, 3]. Since that time, CAD/CAM technology has been used for the fabrication of intracoronal and extracoronal crowns, fixed partial dentures, and implant prostheses. Recently, CAD/CAM technology has been applied to the fabrication of complete dentures [4–12]. By milling complete denture bases from prepolymerized pucks of acrylic resin (AvaDent system, Global Dental Science, Scottsdale, Arizona), the polymerization shrinkage inherent in the traditional methods of fabrication is eliminated. The digital milling process also provides a precise and a reproducible record of the prosthesis design, permitting the fabrication of a duplicate or replacement denture without having to obtain clinical records again.

The digital process of fabricating complete dentures involves scanning conventional complete denture records (described in detail in the previous chapters) [4–12]. It is also possible to scan the neutral zone record [13–16] and the esthetic blueprint (maxillary occlusal rim with the clinically set anterior prosthetic teeth) along with the definitive impression and the interocclusal record, to locate prosthetic teeth positions and to determine the contour and form of the cameo surface of the denture. The CAD/CAM software can also be used for printing record bases and wax trial dentures using stereolithography. Several techniques that can be used to incorporate the neutral zone into the CAD/CAM fabrication of complete dentures are described below.

Registering the Neutral Zone during Impression Making

The neutral zone record can be registered at the time of making definitive impressions [17], using VPS impression material, and then it can be incorporated in the CAD/CAM fabrication of complete dentures.

Technique

1) A conventional clinical impression is made of the edentulous ridges using VPS impression material in stock edentulous impression trays [18]. Note that caution should be exercised to ensure that the thickness of the underlying tray and impression is not excessive, or else registration of the neutral zone will be distorted.

Application of the Neutral Zone in Prosthodontics, First Edition. Joseph J. Massad, David R. Cagna, Charles J. Goodacre, Russell A. Wicks and Swati A. Ahuja.
© 2017 John Wiley & Sons, Inc. Published 2017 by John Wiley & Sons, Inc.
Companion website: www.wiley.com/go/massad/neutral

2) Excess impression material that extends onto the occlusal surface of the impression tray is trimmed away with a scalpel blade ensuring that a minimum of 5 mm of the recorded borders is retained.

3) The occlusal surface is coated with a suitable VPS adhesive and medium-viscosity VPS impression material is applied and extended occlusally and posteriorly up to the level of the center of the retromolar pad (recorded in the definitive impression). Limiting the extension to this height (center of the retromolar pad) permits recording of the neutral zone without having to use an excess amount of VPS material.

4) The impression is accurately seated in the mouth and the patient is asked to swallow three times consecutively while pressing the lips together and then maintaining the lip and tongue positions until the completion of the polymerization of the impression material. Swallowing causes contraction of the lip, cheek, and tongue muscles, with the lateral borders of the tongue producing a depression in the lingual surface of the impression material. Many authors have suggested placing the occlusal plane at the same level as the lateral border of the tongue [19–21]. Thus the occlusal extent of the lingual depression can be used as a guide to determine the level of the occlusal plane (Figure 11.1).

5) A scalpel blade is used to slice through the polymerized impression material horizontally up to the occlusal level of the lingual depression, and the neutral zone is identified and the record is developed by further sculpting of the VPS material (Figure 11.2).

6) The impression is scanned to record both the intaglio surface and the occlusal surface (neutral zone record). A virtual cast (with defined neutral zone) is generated in the software. The opposing arch impression and the interocclusal records are also scanned. A mold of teeth is selected and incorporated into the virtual neutral zone

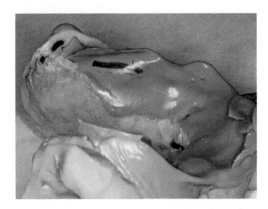

Figure 11.1 Recording the neutral zone at the time of making definitive edentulous ridge impression using medium viscosity VPS impression material. The occlusal extent of the lingual depression marked with a black line, generally denoting the height of the occlusal plane.

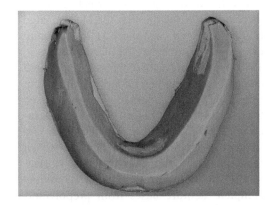

Figure 11.2 Neutral zone impression formed, trimmed occlusally up to the marked black line.

in the software. After establishing the desired occlusion of the prosthetic teeth in the software (AvaDent software, Global Dental Science, Scottsdale, Arizona), trial dentures are milled using the desired shade of tooth-colored resin. After trial placement (Figure 11.3) and making final revisions, the definitive dentures are milled using either a monolithic denture design where the teeth and base are one unit or by bonding the manufacturer's prosthetic teeth into recesses milled in the denture base.

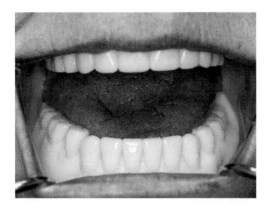

Figure 11.3 Milled trial denture tried in the patient's mouth.

Figure 11.4 Neutral zone impression made using modeling plastic impression compound.

Registering the Neutral Zone during Maxillo-Mandibular Records

The process of registering the neutral zone during maxillo-mandibular records appointment using modeling plastic impression compound is presented in detail in Chapter 7 and will be briefly reviewed here.

Technique

1) The mandibular record base with a modeling plastic impression compound occlusion rim is immersed in a warm water bath set at a temperature of 140 °F and uniformly softened. It is removed from the water bath and quickly placed in the patient's mouth carefully, avoiding distortion [16].
2) The patient is given a glass of warm water and is instructed to swallow, then sip more warm water and swallow again. The sipping and swallowing procedures are repeated several times to mold the compound through the action of the cheek and lip muscles moving inward, and the muscles of the tongue moving outward [16].
3) Once cooled and solidified, the neutral zone record is removed from the mouth, its accuracy is verified, and the excess material is trimmed using a sharp blade (Figure 11.4) [16].

4) The neutral zone record is placed on the cast and it is scanned along with the cast (Figure 11.5a). The esthetic blueprint (EBP) and the opposing cast with interocclusal record are also scanned (Figure 11.5b and c). A three-dimensional electronic image is created of the interarch and neutral zone relationships (Figure 11.5d). Using collective data of the scanned images, denture teeth are digitally planned and positioned to lie within this established space using the planning software program (Figure 11.5e).

Registering the Neutral Zone during the Trial Placement

The neutral zone record can be registered at the trial placement appointment and then be incorporated in the CAD/CAM fabrication of complete dentures.

Technique

1) Conventional complete denture impressions are made and beaded and boxed appropriately, ensuring that the desired amount of the impression borders are visible (Figure 11.6) [22–24]. The boxed impression is then scanned and the scan data is used to print multiple record bases, using stereolithography.

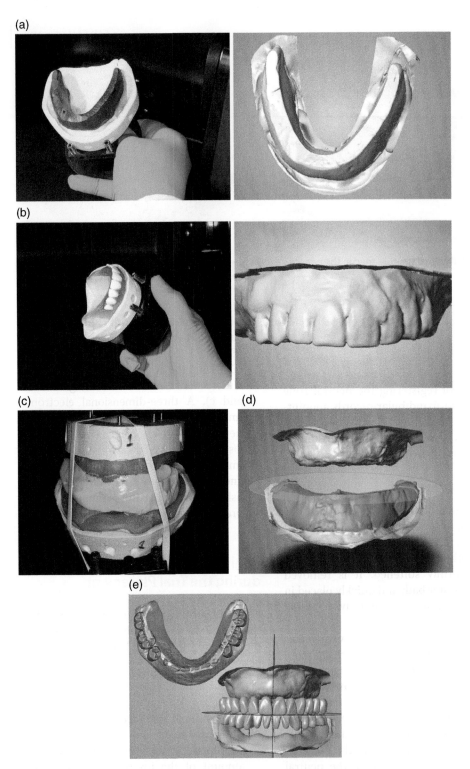

Figure 11.5 (a) Scanning the neutral zone record and mandibular cast (left). Scanned image of the neutral zone record (right); (b) Scanning the EBP wax rim along with the maxillary cast (left). Scanned image of the EBP wax rim (right); (c) scanning the opposing casts with the interocclusal record; (d) a three-dimensional electronic image depicting the interarch relationship and neutral zone space of the mandibular arch; (e) *Top*, neutral zone used to identify prosthetic teeth positions; *Bottom*, maxillary and mandibular teeth digitally arranged.

2) A gothic arch tracing device (Massad Jaw Recorder, Nobilium Company, Albany, NY), is attached to the printed record bases and maxillo-mandibular relationship records are registered (Figure 11.7) [25].

3) An undercontoured wax occlusal rim is added to another printed maxillary record base. Shell teeth (Visionaire Dental Shells, Nobilium Company, Albany, NY) are waxed on the rim and their positions are verified clinically so as to achieve optimal esthetics.

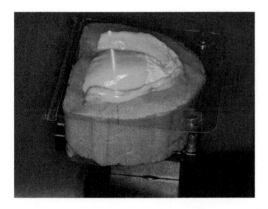

Figure 11.6 Impression boxed to expose the borders.

4) The casts with the maxillo-mandibular records and the wax rim with the shell teeth (EBP) are scanned. The information obtained from the scanned data guides the digital arrangement of the prosthetic teeth. Prosthetic teeth are digitally arranged in the software, and virtual trial dentures are generated and then printed using stereolithography.

5) The trial dentures are used for esthetic evaluation and also to record the neutral zone (termed cameo surface impression) using VPS impression material on the cameo surface of the printed trial denture (Figure 11.8).

6) The clinically verified printed trial dentures with cameo surface impressions are scanned and used to mill definitive dentures (Figure 11.9).

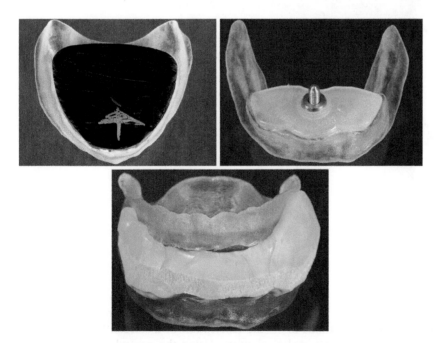

Figure 11.7 Maxillary printed record base with attached tracing plate and completed gothic arch tracing (upper left); mandibular printed record base with central bearing pin assembly attached (upper right); interocclusal record made with gothic arch device (bottom).

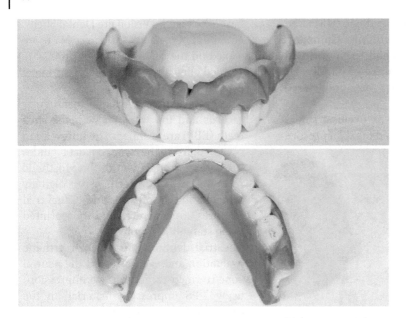

Figure 11.8 Cameo surface neutral zone impressions made on printed trial dentures.

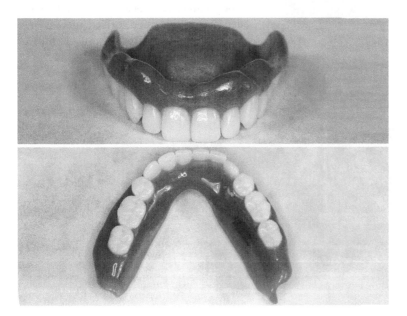

Figure 11.9 Definitive milled dentures.

Summary

This chapter presented various methods of recording the neutral zone, scanning it, and using the resultant digital data for fabricating complete dentures. It also described the process of using neutral-zone scan data for the fabrication of a stereolithographic trial denture. Using the neutral zone as a guide for developing the contours of the polished surface of the mandibular denture and for determining the appropriate bucco-lingual positioning of prosthetic teeth not only aids in achieving retention and stability of the denture but also helps improve speech and patient comfort.

The future will hold advancing applications of technology for the construction of dental prosthetics. The successful use of this technology must not neglect the necessary personal interactions between dentists and their patients.

References

1 M. Andersson, personal communication, 2007.

2 Mörmann, M. H. (2004) The origin of the Cerec method: a personal review of the first 5 years. *Int J Comput Dent*, **7**, 11–24.

3 Mörmann, M. H. (2006) The evolution of the CEREC system. *J Am Dent Assoc*, **137**, suppl., 7S–13S.

4 Kattadiyil, M. T., Goodacre, C. J., and Baba, N. Z. (2013) CAD/CAM Complete dentures: A review of two commercial fabrication systems. *J Calif Dent Assoc*, **41**(6), 407–416.

5 Baba, N. Z., Goodacre, C. J., and Kattadiyil, M. T. (2015) CAD/CAM removable prosthodontics, in *Clinical Applications of Digital Technology* (eds. R. C. Masri, and F. Driscoll), Hoboken, NJ: John Wiley & Sons, Inc.

6 Baba, N. Z. (2016) Materials and processes for CAD/CAM complete denture fabrication. *Curr Oral Health Rep*, doi: 10.1007/s40496-016-0101-3

7 Goodacre, B. J., Goodacre, C. J., Baba, N. Z., and Kattadiyil, M. T. (2016) Comparison of denture base adaptation between CAD/CAM and conventional fabrication techniques. *J Prosthet Dent*, **116**(2), 249–256.

8 Bidra, A. S., Taylor, T. D., and Agar, J. R. (2013) Computer-aided technology for fabricating complete dentures: systematic review of historical background, current status, and future perspectives. *J Prosthet Dent*, **109**(6), 361–366.

9 Bidra, A. S., Farrell, K., Burnham, D. *et al.* (2016) Prospective cohort pilot study of 2-visit CAD/CAM monolithic complete dentures and implant-retained overdentures: clinical and patient-centered outcomes. *J Prosthet Dent*, **115**(5), 578–586.

10 Saponaro, P. C., Yilmaz, B., Heshmati, R. H., and McGlumphy, E. A. (2016) Clinical performance of CAD/CAM-fabricated complete denture: A cross-sectional study. *J Prosthet Dent*. doi: 10.1016/j.prosdent. 2016.03.017 [Epub ahead of print]

11 Saponaro, P. C., Yilmaz, B., Johnston, W., *et al.* (2016) Evaluation of patient experience and satisfaction with CAD/CAM-fabricated complete denture: A retrospective survey study. *J Prosthet Dent*, doi: 10.1016/j. prosdent.2016.01.034 [Epub ahead of print]

12 Kattadiyil, M. T., Jekki, R., Goodacre, C. J., and Baba, N. Z. (2015) Comparison of treatment outcomes in digital and conventional complete removable dental prosthesis fabrications in a predoctoral setting. *J Prosthet Dent*, **114**(6), 818–825.

13 Schiesser, F. J. (1964) The neutral zone and polished surfaces in complete dentures. *J Prosthet Dent*, **14**, 854–865.

14 Beresin, V. E., and Schiesser, F. J. (1976) The neutral zone in complete dentures. *J Prosthet Dent*, **36**, 356–367.

15 Beresin, V. E., and Schiesser, F. J. (eds.) (1979) *Neutral Zone in Complete and Partial Dentures*, 2nd edn. Mosby, St. Louis, MO, pp. 15, 73–108, 158–183.

16 Cagna, D. R., Massad, J. J., and Schiesser, F. J. (2009) The neutral zone revisited: from historical concepts to modern application. *J Prosthet Dent*, **101**, 405–412.

17 Yi-Lin, Y., Yu-Hwa, P., and Ya-Yi, C. (2013) Neutral zone approach to denture fabrication for a severe mandibular ridge resorption patient: Systematic review and modern technique. *J Dent Sci*, **8**, 432–443.

18 Massad, J. J., and Cagna, D. R. (2007) Vinyl polysiloxane impression material in removable prosthodontics. Part 1: edentulous impressions. *Compend Contin Educ Dent*, **28**, 452–460.

19 Yasaki, M. (1961) Height of the occlusion rim and the interocclusal distance. *J Prosthet Dent*, **11**, 26–31.

20 Nagle, R. J., and Sears, V. H. (1962) *Denture Prosthetics*, 2nd edn. Mosby, St. Louis, MO, p. 134.

21 Ghosn, C. A., Zogheib, C., and Makzoume, J. E. (2012) Relationship between the occlusal plane corresponding to the lateral borders of the tongue and the ala-tragus line in edentulous patients. *J Contemp Dent Pract*, **13**, 590–594.

22 Rudd, K. D., Morrow, R. M., and Seldmann, E. E. (1986) Final Impression, boxing and pouring, in *Dental Laboratory Procedures: Vol. 1: Complete Dentures*, 2nd edn. Mosby, St. Louis, MO, pp. 57–79.

23 Powter, R. G., and Hope, M. (1981) A method of boxing impressions. *J Prosthet Dent.*, **45**, 224–225.

24 Bolouri, A., Hilger, T. C., and Gowrylok, M. D. (1975) Boxing impressions. *J Prosthet Dent*, **33**, 692–695.

25 Massad, J. J., Connelly, M. E., Rudd, K. D., and Cagna, D. R. (2004) Occlusal device for diagnostic evaluation of maxillomandibular relationships in edentulous patients: a clinical technique. *J Prosthet Dent*, **91**, 586–590.

Index

Application of the Neutral Zone in Prosthodontics, First Edition. Joseph J. Massad, David R. Cagna,
Charles J. Goodacre, Russell A. Wicks and Swati A. Ahuja.
© 2017 John Wiley & Sons, Inc. Published 2017 by John Wiley & Sons, Inc.
Companion website: www.wiley.com/go/massad/neutral

Printed and bound by CPI Group (UK) Ltd, Croydon, CR0 4YY

27/10/2024

14580245-0002